生态学专业英语教程

主　编　姚晓芹　刘存歧
副主编　楚建周　张艳芬

科学出版社
北京

内 容 简 介

  本书共分为三大部分，第一部分为生态学专业英语基础阅读，选编的文章根据生命系统结构层次进行分类，通过这部分的学习，读者能够掌握该领域的基本词汇和写作方法。第二部分为生态学专业英语提升篇，通过这部分的学习，不仅能够提高读者的阅读能力，而且能够使读者了解生态学研究前沿。第三部分为常用外文数据库介绍及 SCI 论文写作技巧，该部分结合案例详细地介绍了 SCI 论文各部分的写作技巧，并把作者发表论文中遇到的问题和积累的经验一并献给读者。

  本书可作为高等院校生态学相关专业本科生和研究生的专业英语教材，也可供相关研究领域的科研人员阅读。

图书在版编目（CIP）数据

生态学专业英语教程 / 姚晓芹，刘存歧主编. —北京：科学出版社，2017.12

ISBN 978-7-03-053603-7

Ⅰ．①生⋯ Ⅱ．①姚⋯ ②刘⋯ Ⅲ．①生态学-英语-高等学校-教材 Ⅳ．①Q14

中国版本图书馆 CIP 数据核字（2017）第 129416 号

责任编辑：胡云志 滕 云／责任校对：郭瑞芝
责任印制：赵 博／封面设计：华路天然工作室

科 学 出 版 社 出版
北京东黄城根北街 16 号
邮政编码：100717
http://www.sciencep.com

三河市骏杰印刷有限公司印刷
科学出版社发行 各地新华书店经销

*

2017 年 12 月第 一 版　　开本：720×1000　1/16
2024 年 7 月第十次印刷　　印张：14
字数：310 000
定价：52.00 元
（如有印装质量问题，我社负责调换）

前　言

在党的二十大报告中，习近平总书记强调："推动绿色发展，促进人与自然和谐共生。"

生态学是研究生物与环境之间相互关系的一门学科。2011年国务院学位委员会将生态学由二级学科提升为一级学科，充分体现了国家对生态学研究与生态学专业人才培养的高度重视。生态学专业英语是生态学专业课程结构的重要组成部分。为适应近年来生态学学科的飞跃发展和高等学校教学改革的需要，我们着手编写了这本生态学专业英语用书。本书在编写过程中广泛征求学生意见，并依据编者多年生态学专业英语教学实践及英文科技论文撰写和发表经验编写而成，以期能够更好地满足学生对生态学专业英语的需求。

本书共分三大部分，第一部分为生态学专业英语基础阅读和基本专业词汇。所选文章是按照生命系统的结构层次进行分类的，即个体生态、种群生态、群落生态、生态系统生态和全球变化生态。通过这部分内容的学习，读者能够掌握相关领域的基本专业词汇和写作特点。第二部分为生态学专业英语提升篇，这部分所收录的文章都是从与生态学相关的高水平英文期刊上选编而来，并根据需要做了一定的整理和改编。所选论文也是按生命系统的结构层次进行分类的。第三部分为常用外文数据库介绍及SCI论文写作技巧，这部分主要介绍了常见英文数据库的具体使用方法及SCI论文各部分的写作技巧和主要事项，并结合具体案例进行分析。通过这三部分内容的学习使读者在有限的时间内既能掌握一定量的生态学基本专业词汇，又能具备一定的英文科技论文阅读、翻译和写作能力。

在本书编写过程中，河北大学2016级生态学专业研究生郭春延和张楠同学给予了热情的帮助并提出了很好的建议。另外，本书选编了部分同行作者的文稿内容，在此一并向他们表示诚挚的谢意。

本书是"生物学河北省国家重点学科培育项目""河北大学协同育人试点班建设项目（2017）"和"河北省生物工程技术研究中心"系列成果之一，感谢其资金的支持。同时，也感谢科学出版社在本书编写和出版方面给予的大力支持和帮助。

本书初稿已在河北大学 2016 级生态学专业（研究生）的学生中使用，已将教学过程中发现的问题和错误进行了反复的修订，但可能仍存在不足之处，恳请广大读者不吝赐教（yaoxiao301@126.com），以便再版时做得更好。

姚晓芹

2017 年 3 月

2023 年 8 月修改

目 录
Contents

Part Ⅰ Base Components
基 础 篇

Unit 1 Autecology ··· 3

Lesson 1   Biotic and abiotic components ·································· 3
Lesson 2   Soil supports diverse and abundant life ······················· 6
Lesson 3   Plants exhibit adaptations to variations in nutrient availability ····· 10

Unit 2 Population ecology ··· 13

Lesson 4   Population size and density ····································· 13
Lesson 5   Species distribution ··· 16
Lesson 6   Population dynamics ··· 19
Lesson 7   Life history theory ·· 23

Unit 3 Community ecology ·· 26

Lesson 8   The role of species within communities ····················· 26
Lesson 9   Species with a large impact on community structure ····· 29
Lesson 10   Effects of biogeography on community diversity ········· 35
Lesson 11   Ecological succession ··· 38

Unit 4 Ecosystem ecology ··· 41

Lesson 12   Energy flow in ecosystems ····································· 41
Lesson 13   Biogeochemical cycles ·· 45
Lesson 14   Ecosystem dynamics ··· 50
Lesson 15   Ecosystem services ··· 52
Lesson 16   Evolution and natural selection in ecosystems ············· 55

Unit 5　Global change ecology ……………………………………………… 59

Lesson 17　Global change: an overview ……………………………………… 59
Lesson 18　Human impact and ecosystem degradation …………………… 63
Lesson 19　Impacts of global change on human society…………………… 66
Lesson 20　Biodiversity and climate change ……………………………… 69
Lesson 21　Global climate change and risk assessment: invasive species …… 73

Part Ⅱ　Promotion Components
提　升　篇

Unit 6　Studies on autecology ecology …………………………………… 79

Lesson 22　Growth and yield stimulation under elevated CO_2 and drought: a meta-analysis on crops ……………………………………… 79
Lesson 23　Mycorrhizal associations of trees have different indirect effects on organic matter decomposition ……………………………… 81
Lesson 24　Effects of enhanced UV-B radiation on the nutritional and active ingredient contents during the floral development of medicinal chrysanthemum ……………………………………………… 84
Lesson 25　Interactive effects of temperature and pCO_2 on sponges: from the cradle to the grave……………………………………… 86
Lesson 26　Salinity influences arsenic resistance in the xerohalophyte *Atriplex atacamensis* Phil ……………………………………………… 90

Unit 7　Studies on population ecology …………………………………… 93

Lesson 27　Climate, invasive species and land use drive population dynamics of a cold-water specialist ……………………………………… 93
Lesson 28　Does movement behaviour predict population densities? A test with 25 butterfly species……………………………………… 95
Lesson 29　Anthropogenic-driven rapid shifts in tree distribution lead to increased dominance of broadleaf species ……………………… 97
Lesson 30　Species' traits influenced their response to recent climate change …… 99
Lesson 31　Genetic diversity affects the strength of population regulation in a marine fish……………………………………………… 101

目 录

Unit 8　Studies on community ecology ··· **105**

 Lesson 32　Levels and limits in artificial selection of communities ·········· 105

 Lesson 33　Effects of temperature variability on community structure in a natural microbial food web ··· 109

 Lesson 34　Size-balanced community reorganization in response to nutrients and warming ·· 113

 Lesson 35　Functional trait diversity across trophic levels determines herbivore impact on plant community biomass ················· 116

 Lesson 36　Resource pulses can alleviate the biodiversity-invasion relationship in soil microbial communities ·· 119

Unit 9　Studies on ecosystem ecology ·· **122**

 Lesson 37　Energy flows in ecosystems—relationships between predator and prey biomass are remarkably similar in different ecosystems ·· 122

 Lesson 38　Diversity increases carbon storage and tree productivity in Spanish forests ·· 125

 Lesson 39　Biodiversity effects on ecosystem functioning change along environmental stress gradients ··· 128

 Lesson 40　Disturbances catalyze the adaptation of forest ecosystems to changing climate conditions ·· 130

 Lesson 41　Elevated CO_2 and temperature increase soil C losses from a soybean-maize ecosystem ·· 133

Unit 10　Studies on global change ecology ····································· **136**

 Lesson 42　Mycorrhizal status helps explain invasion success of alien plant species ··· 136

 Lesson 43　Genetically informed ecological niche models improve climate change predictions ··· 139

 Lesson 44　Nitrogen deposition and greenhouse gas emissions from grasslands: uncertainties and future directions ··················· 141

 Lesson 45　Climate change is not a major driver of shifts in the geographical distributions of North American birds ······························· 145

 Lesson 46　Temperature impacts on deep-sea biodiversity ····················· 147

Part Ⅲ   An Introduction to Commonly Used Foreign Language Databases and Writing Skills for SCI Papers
常用外文数据库介绍及 SCI 论文写作技巧

Unit 11　SCI 及常用外文数据库介绍 ·· 153
　　Lesson 47　SCI 简介 ··· 153
　　Lesson 48　常用外文数据库介绍 ··· 155
　　Lesson 49　数据库使用方法介绍 ··· 160
　　Lesson 50　百链云服务平台简介 ··· 163

Unit 12　SCI 期刊论文写作技巧 ·· 167
　　Lesson 51　退稿的常见原因 ··· 167
　　Lesson 52　SCI 期刊论文写作前的准备及基本框架 ························· 171
　　Lesson 53　题目、作者及通信地址撰写技巧 ································· 174
　　Lesson 54　摘要和关键词撰写技巧 ·· 177
　　Lesson 55　前言撰写技巧 ·· 179
　　Lesson 56　材料与方法撰写技巧 ··· 182
　　Lesson 57　结果、讨论和结论撰写技巧 ······································· 185
　　Lesson 58　致谢和参考文献撰写技巧 ··· 190
　　Lesson 59　期刊选择技巧 ·· 193
　　Lesson 60　SCI 期刊论文投稿时需要的内容 ·································· 196

References ·· 199

Appendix ··· 205
　　Appendix Ⅰ　常见生态学专业词汇 ·· 205
　　Appendix Ⅱ　生态学学科 SCI 收录期刊简介 ································· 213

Part I Base Components
基 础 篇

The word "ecology" ("Ökologie") was coined in 1866 by the German scientist Ernst Haeckel (1834–1919). Ecology is the scientific analysis and study of interactions among organisms and their environment. Within the discipline of ecology, researchers work at four specific levels that are organism, population, community, and ecosystem, sometimes discretely and sometimes with overlap. Ecosystems are composed of dynamically interacting parts including organisms, the communities they make up, and the non-living components of their environment. The part mainly introduces the basic contents about individual ecology, population ecology, community ecology and ecosystem ecology according to the hierarchical structure of biological systems. In addition, global change ecology is also introduced in the part.

Unit 1 Autecology

In contrast to the study of ecosystems, autecology focuses on individuals and species. Researchers studying ecology at the individual level are interested in the adaptations that enable individuals to live in specific habitats. These adaptations can be morphological, physiological and behavioral.

An important research area within autecology is ecotoxicology how individuals and niches respond to a new biotic or abiotic stress. Such studies not only increase knowledge of the consequences of anthropogenic stresses, but also of how variation plays into evolutionary processes, for instance, how it can enable one species to outcompete another. (http://www.umces.edu/cbl/organismal-ecology)

Lesson 1 Biotic and abiotic components

In <u>biology</u> and <u>ecology</u>, <u>abiotic</u> components or abiotic factors are non-living chemical and physical parts of the <u>environment</u> that affect living organisms and the functioning of <u>ecosystems</u>. Abiotic resources are usually obtained from the <u>lithosphere</u>, <u>atmosphere</u>, and <u>hydrosphere</u>. Examples of abiotic factors are water, air, soil, sunlight, and minerals. Abiotic factors and phenomena associated with them underpin all biology.

<u>Biotic</u> components are the living things in the ecosystem, including <u>producers</u>, <u>consumers</u> and <u>decomposers</u>. A biotic factor is a living component that affects the population of another organism, or the environment. This includes animals that consume the organism, and the

living food that the organism consumes. Biotic factors also include human influence, <u>pathogens</u> and disease outbreaks. Each biotic factor needs energy to do work and food for proper growth.

Relevance

The scope of abiotic and biotic factors spans across the entire <u>biosphere</u>, or global sum of all ecosystems. Such factors can have relevance for an individual within a species, its community or an entire population. For instance, disease is a biotic factor affecting the survival of an individual and its community. Temperature is an abiotic factor with the same relevance.

Some factors have greater relevance for an entire ecosystem. Abiotic and biotic factors combine to create a system, or more precisely, an ecosystem, meaning a community of living and nonliving things considered as a unit. In this case, abiotic factors span as far as the pH of the soil and water, types of nutrients available and even the length of the day. Biotic factors such as the presence of <u>autotrophs</u> or self-nourishing organisms such as plants, and the diversity of consumers also affect an entire ecosystem.

Influencing factors

Abiotic factors affect the ability of organisms to survive and reproduce. Abiotic limiting factors restrict the growth of populations. They help determine the types and numbers of organisms able to exist within an environment.

Biotic factors are living things that directly or indirectly affect organisms within an environment. This includes the organisms themselves, other organisms, interactions between living organisms and even their waste. Other biotic factors include <u>parasitism</u>, disease, and <u>predation</u> (the act of one animal eating another).

Interaction examples

The significance of abiotic and biotic factors comes in their interaction with each other. For a community or an ecosystem to survive, the correct interactions need to be in place.

A simple example would be of abiotic interaction in plants. Water, sunlight and carbon dioxide are necessary for plants to grow. The biotic interaction is that plants use water, sunlight and carbon dioxide to create their own nourishment through a process called <u>photosynthesis</u>.

On a larger scale, abiotic interactions refer to patterns such as climate and seasonality. Factors such as temperature, humidity and the presence or absence of seasons affect the ecosystem. For instance, some ecosystems experience cold winters with a lot of snow. An animal such as a fox within this ecosystem adapts to these abiotic factors by growing a thick, white-colored coat in the winter.

Decomposers such as <u>bacteria</u> and <u>fungi</u> are examples of biotic interactions on such a scale. Decomposers function by breaking down dead organisms. This process returns the basic components of the organisms to the soil, allowing them to be reused within that ecosystem. (http://www.diffen.com/difference/Abiotic_vs_Biotic)

Glossary

biology   生物学
ecology   生态学
abiotic   非生物的
environment   环境
ecosystem   生态系统

lithosphere   岩石圈
atmosphere   大气圈
hydrosphere   水圈
biotic   生物的
producer   生产者

consumer　消费者
decomposer　分解者
pathogen　病原体
biosphere　生物圈
autotroph　自养生物

parasitism　寄生
predation　捕食
photosynthesis　光合作用
bacterium　细菌
fungus　真菌

Lesson 2 Soil supports diverse and abundant life

　　The soil is a radically different environment for life than environments on and above the ground, yet the essential requirements do not differ. Like organisms that live outside the soil, life in the soil requires living space, oxygen, food, and water. Without the presence and intense activity of living organisms, soil development could not proceed. Soil inhabitants from bacteria and fungi to <u>earthworms</u> convert inert mineral matter into a living system.

　　Soil possesses several outstanding characteristics as a medium for life. It is stable structurally and chemically. The soil atmosphere remains saturated or nearly so, until soil moisture drops below a critical point. Soil affords a refuge from extremes in temperature, wind, light, and dryness. These conditions allow soil <u>fauna</u> to make easy adjustments to unfavorable conditions. On the other hand, soil hampers the movement of animals. Except to such channeling species as earthworms, soil pore space is important. It determines the living space, humidity, and gaseous conditions of the soil environment.

　　Only a part of the upper soil layer is available to most soil animals as living space. Spaces within the surface litter, cavities walled off by soil aggregates, pore spaces between individual soil particles, root channels, and fissures are all potential habitats. Most soil animals are limited to pore spaces and cavities larger than themselves.

Unit 1　Autecology

Water in the pore spaces is essential. The majority of soil organisms are active only in water. Soil water is usually present as a thin film coating the surfaces of soil particles. This film contains, among other things, bacteria, <u>unicellular algae</u>, <u>protozoa</u>, <u>rotifers</u>, and <u>nematodes</u>. The thickness and shape of the water film restrict the movement of most of these soil organisms. Many small species and immature stages of larger <u>centipedes</u> and <u>millipedes</u> are immobilized by a film of water and are unable to overcome the surface tension imprisoning them. Some soil animals, such as millipedes and centipedes, are highly susceptible to desiccation and avoid it by burrowing deeper.

When water fills pore spaces after heavy rains, conditions are disastrous for some soil inhabitants. If earthworms cannot evade flooding by digging deeper, they come to the surface, where they often die from ultraviolet radiation and desiccation or are eaten.

A diversity of life occupies these habitats. The number of species of bacteria, fungi, protists, and representatives of nearly every invertebrate phylum found in the soil is enormous. A soil zoologist found 110 species of <u>beetles</u>, 229 species of <u>mites</u>, and 46 species of snails and slugs in the soil of an Austrian deciduous beech forest.

Dominant among the soil organisms are bacteria, fungi, protozoans, and nematodes. Flagellated protozoans range from 100,000 to 1,000,000, amoebas from 50,000 to 500,000, and ciliates up to 1,000 per gram of soil. Nematodes occur in the millions per square meter of soil. These organisms obtain their nourishment from the roots of living plants and from dead organic matter. Some protozoans and free-living nematodes feed selectively on bacteria and fungi.

Living within the pore spaces of the soil are the most abundant and widely distributed of all forest soil animals, the mites (Acarina) and <u>springtails</u> (Collembola). Together they make up over 80 percent of the animals in the soil. Flattened, they wiggle, squeeze, and digest their way through tiny caverns in the soil. They feed on fungi or search for prey in the dark interstices and pores of the organic and mineral mass.

The more numerous of the two, both in species and numbers, are the mites, tiny eight-legged <u>arthropods</u> from 0.1 to 2.0 mm in size. The most common mites in the soil and litter are the Orbatei. They live mostly on fungal <u>hyphae</u> that attack dead vegetation as well as on the sugars produced by the digestion of <u>cellulose</u> found in conifer needles.

Collembola are the most widely distributed of all insects. Their common name, springtail, describes the remarkable springing organ at the posterior end, which enables them to leap great distances for their size. The springtails are small, from 0.3 to 1.0 mm in size. They consume decomposing plant material, largely for the fungal hyphae they contain.

Prominent among the larger soil fauna are the earthworms (Lumbricidae). <u>Burrowing</u> through the soil, they ingest soil and fresh litter and egest both mixed with intestinal secretions. Earthworms defecate aggregated castings on or near the surface of the soil or as a semiliquid in intersoil spaces along the burrow. These aggregates produce a more open structure in heavy soil and bind light soil together. In this manner, earthworms improve the soil environment for other organisms.

Feeding on the surface litter are millipedes. They eat leaves, particularly those somewhat decomposed by fungi. Lacking the enzymes necessary for the breakdown of cellulose, millipedes live on the fungi contained within the litter. The millipedes' chief contribution is the mechanical breakdown of litter, making it more vulnerable to microbial attack, especially by <u>saprophytic</u> fungi.

Accompanying the millipedes are snails and slugs. Among: the soil invertebrates, they possess the widest range of enzymes to hydrolyze cellulose and other plant <u>polysaccharides</u>, possibly even the highly indigestible <u>lignins</u>.

Not to be ignored are <u>termites</u> (Isoptera), white, wingless, social insects. Except for some <u>dipteran</u> and <u>beetle larvae</u>, termites are the only larger soil inhabitants that can break down the cellulose of wood. They do so with the aid of symbiotic protozoans living in their gut. Termites dominate the tropical soil fauna. In the tropics termites are responsible for the rapid removal of wood, dry grass, and other materials from the soil surface. In constructing their huge and complex mounds, termites move considerable amounts of soil.

Detrital-feeding organisms support predators. Small arthropods are the principal prey of <u>spiders</u>, beetles, <u>pseudoscorpions</u>, predaceous mites, and centipedes. Protozoans, rotifers, myxobacteria, and nematodes feed on bacteria and algae. Various predaceous fungi live on bacteria-feeders and algae-feeders. (Smith and Smith, 2003)

Glossary

earthworm　蚯蚓
fauna　动物群
unicellular algae　单细胞藻类
protozoa　原生动物
rotifer　轮虫类
nematode　线虫类
centipede　蜈蚣
millipede　千足虫
beetle　甲壳虫
mite　螨
springtail　跳虫
arthropod　节肢动物

hyphae　菌丝
cellulose　纤维素
burrow　地洞
saprophytic　腐生的
polysaccharide　多糖
lignin　木质素
termite　白蚁
dipteran　双翅昆虫
beetle larvae　甲虫幼虫
spider　蜘蛛
pseudoscorpion　拟蝎

Lesson 3 Plants exhibit adaptations to variations in nutrient availability

Plants require 16 elements to carry out their metabolic processes and to synthesize new <u>tissues</u>. Thus, the availability of nutrients has many direct effects on plant survival, growth, and <u>reproduction</u>. Of the <u>macronutrients</u>, carbon (C), hydrogen (H), and oxygen (O) form the majority of plant tissues. These elements are derived from CO_2 and H_2O and are made available to the plant as <u>glucose</u> through photosynthesis. The remaining 6 macronutrients—nitrogen (N), phosphorus (P), potassium (K), calcium (Ca), magnesium (Mg), and sulfur (S)—exist in a variety of states in the soil, and their availability to plants is affected by several important and different processes. In the case of terrestrial environments, plants take up nutrients from the soil. In aquatic environments, plants take up nutrients from the substrate or directly from the water.

The best example of the direct link between nutrient availability and plant performance involves nitrogen. Nitrogen plays a major role in photosynthesis. There are two important compounds in photosynthesis—the enzyme <u>rubisco</u> and the pigment <u>chlorophyll</u>. Rubisco catalyzes the transformation of carbon dioxide into simple sugars, and chlorophyll absorbs light energy. Nitrogen is a component of both compounds; the plant requires nitrogen to make them. In fact, over 50 percent of the nitrogen content of a leaf is in some way involved directly with the process of photosynthesis, with much of it tied up in these two compounds. As a result, the (maximum) rate of photosynthesis for a species is correlated with the nitrogen content of its leaves.

The uptake of a nutrient depends upon both supply and demand. Low concentrations in the soil or water mean low uptake rates. A lower uptake rate decreases the concentrations of nitrogen in the leaf and, consequently, the concentrations of rubisco and chlorophyll. Therefore, lack of nitrogen limits the growth of plants. Geology, climate, and biological activity alter the availability of nutrients in the soil. As a consequence, some environments are rich in nutrients, while others are poor. How do plants from low-nutrient environments succeed?

Because plants require nutrients for the synthesis of new tissue, a plant's rate of growth influences its demand for a nutrient. In turn, the plant's uptake rate of the nutrient also influences growth. This relationship may seem circular, but the important point is that not all plants have the same inherent (maximum potential) rate of growth. For example, shade-tolerant plants have an inherently lower rate of photosynthesis and growth than shade-intolerant plants, even under high light conditions. This lower rate of photosynthesis and growth means a lower demand for resources, including nutrients. The same pattern of reduced photosynthesis occurs among plants that are characteristic of low-nutrient environments. Bradshaw et al. reported the growth responses of two grass species when soil is enriched with nitrogen. The species that naturally grows in a high-nitrogen environment keeps increasing its rate of growth with increasing nitrogen. The species native to a low-nitrogen environment reaches its maximum rate of growth at low to medium nitrogen availability. It does not respond to further additions of nitrogen.

Some plant ecologists suggest that a low natural growth rate is an adaptation to a low-nutrient environment. One advantage of slower growth is that the plant can avoid stress under low-nutrient conditions. A slow-growing plant can still maintain optimal rates of photosynthesis and other processes critical for growth under low-nutrient availability. In contrast, a plant with an inherently high rate of growth will show signs of stress.

A second adaptation to low-nutrient environments is leaf <u>longevity</u>. The production of a leaf has a cost to the plant. This cost can be defined in terms of the carbon and other nutrients required to make the leaf. At a low rate of photosynthesis, a leaf needs a longer time to "pay back" the cost of its production. As a result, plants inhabiting low-nutrient environments tend to have longer-lived leaves. A good example is the dominance of pine species on poor, sandy soils in the coastal region of the southeastern United States. In contrast to <u>deciduous</u> tree species, which shed their leaves every year, pines have needles that live for tip to 3 years.

Like water, nutrients are a below-ground resource of terrestrial plants. Their ability to exploit this resource is related to the amount of root mass. One means by which plants growing in low-nutrient environments compensate is to increase the production of roots. This increase is one cause of their low growth rates. Just as was the case with water limitation, carbon is allocated to the production of roots at the cost of the production of leaves. The reduced leaf area reduces the rate of carbon uptake in photosynthesis relative to the rate of carbon loss in <u>respiration</u>. The result is a lower net carbon gain and growth rate by the plant.（Smith and Smith, 2003）

Glossary

tissue　组织
reproduction　繁殖
macronutrient　大量元素
glucose　葡萄糖
rubisco　核酮糖-1,5-二磷酸羧化/加氧酶

chlorophyll　叶绿素
geology　地质学，地质情况
inherent　内在的，天生的
longevity　寿命
deciduous　每年落叶的
respiration　呼吸

Unit 2   Population ecology

Population ecology is a sub-field of ecology that deals with the dynamics of species populations and how these populations interact with the environment. It is the study of how the population sizes of species change over time and space. The term population ecology is often used interchangeably with population biology or population dynamics. The unit mainly introduces population size and density, species distribution, population dynamics and life history theory.

Lesson 4   Population size and density

Population size and density are the two most important statistics scientists use to describe and understand populations. A population's size refers to the number of individuals it comprises. Its density is the number of individuals within a given area or volume. These data allow scientists to model the fluctuations of a population over time. For example, a larger population may be more stable than a smaller population. With less genetic variation, a smaller population will have reduced capacity to adapt to environmental changes. Individuals in a low-density population are thinly dispersed; hence, they may have more difficulty finding a mate compared to individuals in a higher-density population. On the other hand, high-density populations often result in increased competition for food. Many factors influence density, but, as a rule-of-thumb, smaller organisms have higher population densities than do larger organisms.

Counting all individuals in a population is the most accurate way to determine its size. However, this approach is not usually feasible, especially for large populations or extensive habitats. Instead, scientists study populations by sampling, which involves counting individuals within a certain area or volume that is part of the population's habitat. Analyses of sample data enable scientists to infer population size and population density about the entire population.

A variety of methods can be used to sample populations. Scientists usually estimate the populations of sessile or slow-moving organisms with the quadrat method. A quadrat is a square that encloses an area within a habitat. The area may be defined by staking it out with sticks and string, or using a square made of wood, plastic, or metal placed on the ground.

A field study usually includes several quadrat samples at random locations or along a transect in representative habitat. After they place the quadrats, researchers count the number of individuals that lie within the quadrat boundaries. The researcher decides the quadrat size and number of samples from the type of organism, its spatial distribution, and other factors. For sampling daffodils, a 1 m^2 quadrat could be appropriate. Giant redwoods are larger and live further apart from each other, so a larger quadrat, such as 100 m^2, would be necessary. The correct quadrat size ensures counts of enough individuals to get a sample representative of the entire habitat.

Scientists typically use the mark and recapture technique for mobile organisms such as mammals, birds, or fish. With this method, researchers capture animals and mark them with tags, bands, paint, body markings, or some other sign. The marked animals are then released back into their environment where they mix with the rest of the population. Later, a new sample is collected, including some individuals that are marked (recaptures) and some individuals that are unmarked.

The ratio of marked to unmarked individuals allows scientists to

Unit 2 Population ecology

calculate how many individuals are in the population as an estimate of total population size. This method assumes that the larger the population, the lower the percentage of tagged organisms that will be recaptured since they will have mixed with more untagged individuals.

The mark and recapture method has limitations. Some animals from the first catch may learn to avoid capture in the second round. Such behavior would cause inflated population estimates. Alternatively, animals may preferentially be retrapped (especially if a food reward is offered), resulting in an underestimate of population size. Also, some species may be harmed by the marking technique, reducing their survival. A variety of other techniques have been developed, including the electronic tracking of animals tagged with radio transmitters and the use of data from commercial fishing and trapping operations to estimate the size and health of populations and communities. （https://www.boundless.com/biology/textbooks/boundless-biology-textbook/population-and-community-ecology-45/population-demography-249/population-size-and-density-925-12181/）

Glossary

population 种群
individual 个体
fluctuation 波动
genetic 遗传的，基因的，起源的
rule-of-thumb 经验法则
habitat （动物的）栖息地，生境

quadrat 样方
organism 生物体，有机体
distribution 分配，分布
recapture technique 重捕技术
mammal 哺乳动物
survival 幸存，生存

Lesson 5　Species distribution

Density and size are useful measures for characterizing populations. Scientists gain additional insight into a species' biology and ecology from studying how individuals are spatially distributed. Dispersion or distribution patterns show the spatial relationship between members of a population within a habitat. Patterns are often characteristic of a particular species; they depend on local environmental conditions and the species' growth characteristics (as for plants) or behavior (as for animals).

Individuals of a population can be distributed in one of three basic patterns: they can be more or less equally spaced apart (uniform dispersion), dispersed randomly with no predictable pattern (random dispersion), or clustered in groups (clumped dispersion).

Clumped distribution

Clumped distribution is the most common type of dispersion found in nature. In clumped distribution, the distance between neighboring individuals is minimized. This type of distribution is found in environments that are characterized by patchy resources. Animals need certain resources to survive, and when these resources become rare during certain parts of the year animals tend to "clump" together around these crucial resources. Individuals might be clustered together in an area due to social factors such as selfish herds and family groups. Organisms that usually serve as prey form clumped distributions in areas where they can hide and detect predators easily.

Regular or uniform distribution

Less common than clumped distribution, uniform distribution, also known as even distribution, is evenly spaced. Uniform distributions are found in populations in which the distance between neighboring individuals is maximized. The need to maximize the space between individuals generally arises from competition for a resource such as moisture or nutrients, or as a result of direct social interactions between individuals within the population, such as territoriality. For example, penguins often exhibit uniform spacing by aggressively defending their territory among their neighbors. The burrows of great gerbils for example are also regularly distributed, which can be seen on satellite images. Plants also exhibit uniform distributions, like the creosote bushes in the southwestern region of the United States. *Salvia leucophylla* is a species in California that naturally grows in uniform spacing. This flower releases chemicals called terpenes which inhibit the growth of other plants around it and results in uniform distribution. This is an example of allelopathy, which is the release of chemicals from plant parts by leaching, root exudation, volatilization, residue decomposition and other processes. Allelopathy can have beneficial, harmful, or neutral effects on surrounding organisms. Some allelochemicals even have selective effects on surrounding organisms; for example, the tree species *Leucaena leucocephala* exudes a chemical that inhibits the growth of other plants but not those of its own species, and thus can affect the distribution of specific rival species. Allelopathy usually results in uniform distributions, and its potential to suppress weeds is being researched. Farming and agricultural practices often create uniform distribution in areas where it would not previously exist, for example, orange trees growing in rows on a plantation.

Random distribution

Random distribution, also known as unpredictable spacing, is the least common form of distribution in nature and occurs when the members of a given species are found in environments in which the position of each individual is independent of the other individuals: they neither attract nor repel one another. Random distribution is rare in nature as biotic factors, such as the interactions with neighboring individuals, and abiotic factors, such as climate or soil conditions, generally cause organisms to be either clustered or spread. Random distribution usually occurs in habitats where environmental conditions and resources are consistent. This pattern of dispersion is characterized by the lack of any strong social interactions between species. For example, when <u>dandelion</u> seeds are dispersed by wind, random distribution will often occur as the seedlings land in random places determined by uncontrollable factors. <u>Oyster larvae</u> can also travel hundreds of kilometers powered by sea currents, which can result in their random distribution.

Abiotic and biotic factors

The distribution of species into clumped, uniform, or random depends on different abiotic and biotic factors. Any non-living chemical or physical factor in the environment is considered an <u>abiotic factor</u>. There are three main types of abiotic factors: climatic factors consist of sunlight, atmosphere, humidity, temperature, and salinity; <u>edaphic factors</u> are abiotic factors regarding soil, such as the coarseness of soil, local geology, soil pH, and aeration; and social factors include land use and water availability. An example of the effects of abiotic factors on species distribution can be seen in drier areas, where most individuals of a species will gather around water sources, forming a clumped distribution.

<u>Biotic factors</u>, such as predation, disease, and competition for resources such as food, water, and mates, can also affect how a species is distributed. A biotic factor is any behavior of an organism that affects

another organism, such as a predator consuming its prey. For example, biotic factors in a quail's environment would include their prey (insects and seeds), competition from other quail, and their predators, such as the coyote. An advantage of a herd, community, or other clumped distribution allows a population to detect predators earlier, at a greater distance, and potentially mount an effective defense. Due to limited resources, populations may be evenly distributed to minimize competition, as is found in forests, where competition for sunlight produces an even distribution of trees. (https://en.wikipedia.org/wiki/Species_distribution)

Glossary

uniform dispersion　均匀分布
random dispersion　随机分布
clumped dispersion　集群分布
resource　资源
predator　食肉动物
competition　生存竞争
penguin　企鹅
gerbil　沙鼠
terpene　萜烯

allelopathy　化感作用
root exudation　根系分泌物
allelochemical　化感物质
dandelion　蒲公英
oyster larvae　牡蛎幼虫
abiotic factor　非生物因素
edaphic factor　土壤因子
biotic factor　生物因素

Lesson 6　Population dynamics

Population size, density, and distribution patterns describe a population at a fixed point in time. To study how a population changes over time, scientists must use the tools of <u>demography</u>: the statistical study of population changes over time. The key statistics demographers use are <u>birth rates</u>, <u>death rates</u>, and <u>life expectancies</u>; although, in practice, scientists also study <u>immigration</u> and <u>emigration</u> rates, which also affect populations.

These measures, especially birth rates, may be related to the population characteristics. For example, a large population would have a relatively-high birth rate if it has more reproductive individuals. Alternatively, a large population may also have a high death rate because of competition, disease, or waste accumulation. A high population density may lead to more reproductive encounters between individuals, as would a clumped distribution pattern. Such conditions would increase the birth rate.

Biological features of the population also affect population changes over time. Birth rates will be higher in a population with the ratio of <u>males</u> to <u>females</u> biased towards females, or in a population composed of relatively more individuals of reproductive age.

The demographic characteristics of a population are the basic determinants of how the population changes over time. If birth and death rates are equal, the population remains stable. The population will increase if birth rates exceed death rates, but will decrease if birth rates are lower than death rates. Life expectancy, another important factor, is the length of time individuals remain in the population. It is impacted by local resources, <u>reproduction</u>, and the overall health of the population. These demographic characteristics are often displayed in the form of a <u>life table</u>.

Life tables

Life tables, which provide important information about the life history of an organism, divide the population into age groups and often sexes; they show how long a member of that group will probably live. The tables are modeled after actuarial tables used by the insurance industry for estimating human life expectancy. Life tables may include: the probability of individuals dying before their next birthday (i.e., mortality rate); the percentage of surviving individuals at a particular age interval; the life expectancy at each interval.

There are two types of life tables used in actuarial science. The period life table represents mortality rates during a specific time period of a certain population. A cohort life table, often referred to as a generation life table, is used to represent the overall mortality rates of a certain population's entire lifetime. They must have had to be born during the same specific time interval. A cohort life table is more frequently used because it is able to make a prediction of any expected changes in mortality rates of a population in the future. This type of table also analyzes patterns in mortality rates that can be observed over time. Both of these types of life tables are created based on an actual population from the present, as well as an educated prediction of the experience of a population in the near future.

Survivorship curves

Another tool used by population ecologists is a <u>survivorship curve</u>, which is a graph of the number of individuals surviving at each age interval plotted versus time (usually with data compiled from a life table). These curves allow comparison of <u>life histories</u> of different populations.

Humans and most primates exhibit a Type I survivorship curve because a high percentage of offspring survive early and middle years; death occurs predominantly in older individuals. These species have few <u>offspring</u> as they invest in parental care to increase survival.

Birds show the Type II survivorship curve because equal numbers of birds tend to die at each age interval. These species may also have relatively-few offspring and provide significant parental care.

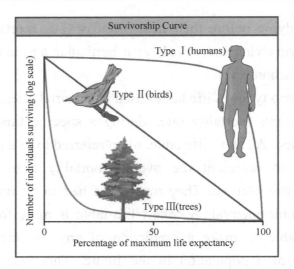

Trees, marine invertebrates, and most fishes exhibit a Type III survivorship curve. Very few individuals survive the younger years; however, those that live to old age are likely to survive for a relatively-long period. Organisms in this category usually have large numbers of offspring and provide little parental care. Such offspring are "on their own" and suffer high mortality due to predation or starvation; however, their abundance ensures that enough individuals survive to the next generation, perpetuating the population. （https://www.boundless.com/biology/textbooks/boundless-biology-textbook/population-and-community-ecology-45/population-demography-249/the-study-of-population-dynamics-927-12183/）

Glossary

demography　人口统计学
birth rate　出生率
death rate　死亡率
life expectancy　预期寿命

immigration　迁入
emigration　迁出
male　雄性
female　雌性

life table　生命表　　　　　　　life history　生活史
survivorship curve　存活曲线　　offspring　后代，子代

Lesson 7　Life history theory

　　Life history theory is a theory of biological evolution that seeks to explain aspects of organisms' <u>anatomy</u> and behavior by reference to the way that their life histories—including their reproductive development and behaviors, life span and post-reproductive behavior—have been shaped by <u>natural selection</u>. These events, notably juvenile development, age of sexual maturity, first reproduction, number of offspring and level of parental investment, senescence and death, depend on the physical and ecological environment of the organism.

　　<u>Reproductive strategies</u> play a key role in life histories, they do not account for important factors such as limited resources and competition. The regulation of population growth by these factors can be used to introduce a classical concept in population biology: r/K selection theory. In ecology, r/K selection theory relates to the selection of combinations of traits in an organism that trade off between quantity and quality of offspring. The focus upon either increased quantity of offspring at the expense of individual parental investment of r-strategists, or reduced quantity of offspring with a corresponding increased parental investment of K-strategists, varies widely, seemingly to promote success in particular environments.

K-selected species

　　K-selected species are those in stable, predictable environments. Populations of K-selected species tend to exist close to their carrying capacity (hence the term K-selected) where <u>intraspecific competition</u> is high. These species produce few offspring, have a long gestation period, and often give long-term care to their offspring. While larger in size when born, the offspring are relatively helpless and immature at birth. By the time they reach adulthood, they must develop skills to compete for natural resources. Examples of *K*-selected species are primates including humans, other mammals such as elephants, and plants such as oak trees.

***r*-selected species**

In contrast to *K*-selected species, *r*-selected species have a large number of small offspring (hence their *r* designation). This strategy is often employed in unpredictable or changing environments. Animals that are *r*-selected do not give long-term parental care and the offspring are relatively mature and self-sufficient at birth. Examples of *r*-selected species are marine invertebrates, such as jellyfish, and plants, such as the dandelion. Dandelions have small seeds that are dispersed long distances by wind; many seeds are produced simultaneously to ensure that at least some of them reach a hospitable environment. Seeds that land in inhospitable environments have little chance for survival since the seeds are low in energy content. Note that survival is not necessarily a function of energy stored in the seed itself.

Modern theories of life history

The *r*- and *K*-selection theory, although accepted for decades and used for much groundbreaking research, has now been reconsidered. Many population biologists have abandoned or modified it. Over the years, several studies attempted to confirm the theory, but these attempts have largely failed. Many species were identified that did not follow the theory's predictions. Furthermore, the theory ignored the age-specific mortality of the populations which scientists now know is very important.

New demographic-based models of life history evolution have been developed which incorporate many ecological concepts included in *r*-and *K*-selection theory, as well as population age structure and mortality factors.（https://www.boundless.com/biology/textbooks/boundless-biology-textbook/population-and-community-ecology-45/life-history-patterns-250/theories-of-life-history-932-12189/）

Glossary

anatomy　解剖
natural selection　自然选择
reproductive strategy　生殖策略
intraspecific competition　种内竞争
gestation　孕期

jellyfish　水母
dandelion　蒲公英
hospitable　（气候，环境）宜人的，适宜居住的
inhospitable　荒凉的, 不适宜居住的

Unit 3   Community ecology

In ecology, a community or biocoenosis is an assemblage or association of populations of two or more different species occupying the same geographical area in a particular time.

Community ecology or synecology is the study of the interactions between species in communities on many spatial and temporal scales, including the distribution, structure, abundance, demography, and interactions between coexisting populations. The primary focus of community ecology is on the interactions between populations, as is determined by specific genotypic and phenotypic characteristics. Modern community ecology examines patterns such as variation in species richness, equitability, productivity and food web structure; it also examines processes such as predator-prey population dynamics, succession, and community assembly.

Lesson 8   The role of species within communities

Communities are complex entities that can be characterized by their structure (the types and numbers of species present) and dynamics (how communities change over time). Understanding community structure and dynamics enables community ecologists to manage ecosystems more effectively. There are three main types of species that serve as the basis for a community. These include the foundation species, keystone species, and invasive species. Each of these has a specific role in how communities are formed.

Keystone species

A keystone species is a species that has a disproportionately large effect on its environment relative to its abundance. Such species are described as playing a critical role in maintaining the structure of an ecological community, affecting many other organisms in an ecosystem and helping to determine the types and numbers of various other species in the community. An ecosystem may experience a dramatic shift if a keystone species is removed, even though that species was a small part of the ecosystem by measures of biomass or productivity. It became a popular concept in conservation biology. Although the concept is valued as a descriptor for particularly strong inter-species interactions, and it has allowed easier communication between ecologists and conservation policy-makers, it has been criticized for oversimplifying complex ecological systems.

Foundation species

In ecology, the term foundation species is used to refer to a species that has a strong role in structuring a community. A foundation species can occupy any trophic level in a food web (i.e., they can be primary producers, herbivores or predators). The term was coined by Paul K. Dayton in 1972, who applied it to certain members of marine invertebrate and algae communities. It was clear from studies in several locations that there were a small handful of species whose activities had a disproportionate effect on the rest of the marine community and they were therefore key to the resilience of the community. Dayton's view was that focusing on foundation species would allow for a simplified approach to more rapidly understand how a community as a whole would react to disturbances, such as pollution, instead of attempting the extremely difficult task of tracking the responses of all community members simultaneously. The term has since been applied to range of organisms in ecosystems around the world, in both aquatic and terrestrial environments. Aaron Ellison et al. introduced the term to <u>terrestrial ecology</u> by applying the term foundation species to tree species that define and structure certain forest ecosystems through their influences on associated organisms and modulation of ecosystem processes.

Invasive species

Invasive species are foreign species whose introduction can cause harm to the economy and the environment. Invasive species are often better competitors than native species, resulting in population explosions. These new species usually overtake the native populations, driving them to localized <u>extinctions</u>.

While all species compete to survive, invasive species appear to have specific traits or specific combinations of traits that allow them to outcompete <u>native species</u>. In some cases, the competition is about rates of growth and reproduction. In other cases, species interact with each other more directly.

Researchers disagree about the usefulness of traits as invasiveness markers. One study found that of a list of invasive and noninvasive species, 86% of the invasive species could be identified from the traits alone. Another study found invasive species tended to have only a small subset of the presumed traits and that many similar traits were found in noninvasive species, requiring other explanations. Common invasive species traits include the following: Fast growth, Rapid reproduction, High dispersal ability, <u>Phenotypic plasticity</u> (the ability to alter growth form to suit current conditions), <u>Tolerance</u> of a wide range of environmental conditions (Ecological competence), Ability to live off of a wide range of food types (generalist), Association with humans, and Prior successful invasions.

Unit 3 Community ecology

(https://en.wikipedia.org/wiki/Keystone_species; https://en.wikipedia.org/wiki/ Foundation_species; https://www. boundless .com/biology/）

Glossary

community　群落
dynamics　动力学
ecosystem　生态系统
keystone species　关键物种
invasive species　入侵物种
abundance　丰度
biomass　生物量
productivity　生产力
trophic level　营养级

food web　食物网
producer　生产者
herbivore　食草动物
predator　食肉动物
terrestrial ecology　陆地生态学
extinction　灭绝
native species　本地物种
phenotypic plasticity　表型可塑性
tolerance　耐受性

Lesson 9　Species with a large impact on community structure

Dominant species are the most abundant species in a community, exerting a strong influence over the occurrence and distribution of other species. In contrast, keystone species have effects on communities that far exceed their abundance. That is to say, the importance of keystone species would not be predicted based upon their occurrence in an ecosystem. Dominant and keystone species influence the presence and abundance of other organisms through their feeding relationships. Feeding relationships—eating or being eaten—are called trophic interactions.

In addition, some organisms, called foundation species, exert influence on a community not through their trophic interactions, but by causing physical changes in the environment. These organisms alter the environment through their behavior or their large collective biomass. Foundation species may also be dominant species.

Predation can have large effects on prey populations and on community structure. Predators can increase diversity in communities by preying on competitive dominant species or by reducing consumer pressure on foundation species. For example, in rocky intertidal systems of the Pacific Northwestern US, mussels, barnacles, and seaweeds require a hard substrate to grow on, and they compete for space on the rocks. Mussels (dominant species) are superior competitors and can exclude all other species within a few years. However, starfish (keystone species) preferentially consume mussels, and in doing so, free up space for many other organisms to settle and grow, thus increasing biodiversity within this ecosystem. Similarly, kelp forests in Alaska are home to numerous species of fish and invertebrates, but these giant kelps, which are the dominant and foundation species of kelp forest communities, can be completely destroyed by sea urchins grazing. Urchins consume the kelp and create barren areas devoid of life. Urchins however are readily consumed by sea otters (keystone species), and by keeping urchin numbers low, otters assure that the kelp forest community remains intact.

Communities can be structured by 'bottom-up' or 'top-down' forces

Bottom-up forces influence communities from lower to higher trophic levels of the food chain. For example, if nutrient levels rise, stimulating the growth of vegetation, then the higher trophic levels should also increase in biomass in a community structured through bottom-up mechanisms. Hawaiian forests are often nutrient limited, and when nutrients are added, vegetation increases as do higher trophic levels.

Predation is a top-down force because the effects of predators start at the top of the food chain and cascade downward to lower trophic levels. A <u>trophic cascade</u> occurs when predators indirectly affect the abundance of organisms more than two trophic levels down.

The otter-urchin-kelp interaction is an example of a trophic cascade. In this system, the otter (keystone species) increases the abundance of kelp (foundation species) by consuming urchins, thereby decreasing urchin grazing on kelp. Another example of how top-down forces affect communities via trophic cascades can be found in <u>Yellowstone National Park</u>, USA. A program to reintroduce wolves to Yellowstone has led to an increase in vegetation because wolves (top-predator) consume elk (intermediate consumer), one of the primary grazers in the park. Thus, top-down control (i.e., consumption) of elk by wolves alleviates grazing by elk and increases the abundance of <u>primary producers</u>. Similarly, predation by spiders on grasshoppers decreases grazing on plants in the fields of New England and predation by <u>planktivorous fish</u> on <u>zooplankton</u> increases the abundance of <u>phytoplankton</u> in freshwater lakes.

Human activities adversely affect communities by removing important species, especially predators

These and many similar observations suggest that predators play key roles in determining the presence and abundance of many species aquatic and terrestrial communities. Unfortunately, human activities are causing the populations of many predatory species to decline worldwide. These declines may have significant consequences for communities, and <u>deprive</u> humans of the benefits we receive from these natural communities. In coastal systems, scallops and other bivalves are consumed by stingrays, which in turn are preyed upon by sharks. <u>Overfishing</u> of large shark species (top-predators) has led to an increase in the numbers of rays (intermediate consumers), and greater predation by sting rays has destroyed the scallop fishery along the East Coast of the US. Likewise, in Alaska, sea otter decline has led to an increase in sea urchin abundance and a loss of kelp forest, and this decline has been attributed to greater predation on sea otters by killer whales. Killer whales did not begin eating otters until their preferred prey, sea lions, became less abundant. The decline in sea lion populations likely resulted from overfishing of pollock. Pollock are fish and are sea lions' primary food source. Thus, overfishing of pollock led a decline in sea lion populations, causing killer whales to seek an alternative food source (otters). The change in predation by killer whales removed an important predator in coastal Alaska and resulted in the loss of kelp forest <u>habitat</u>. These are just two examples of how human activities can have large, unintended consequences on the composition of entire <u>biological communities</u>.

Predators may affect communities through lethal and non-lethal processes

<u>Lethal effect</u> (sometimes referred to as a consumptive effect) occurs when predators consume lower trophic levels. <u>Non-lethal effect</u> (also referred to as a non-consumptive effect) occurs when prey react to predators by altering their behavior, morphology, and/or habitat selection. Classic studies of predation, such as those described above, have focused on the lethal or consumptive effects predators have on lower trophic levels. That is to say, predators consume prey, and by reducing prey numbers, have cascading and sometimes large effects upon communities. Recent studies however have shown that predators also affect prey populations through non-lethal or non-consumptive means. In these situations, predators alter prey behavior, morphology, and/or habitat selection. Some prey species may remain in refuges and forgo foraging opportunities to avoid predators, while others may alter their morphology to make

Unit 3 Community ecology

themselves less susceptible to predation. Changes in behavior or morphology are often necessary to minimize predation risk, but are costly to prey resulting in decreased growth and fecundity. Examples of non-lethal predator effects are numerous, and have recently been shown to affect community composition in much the same way lethal predator effects do. That is, a trophic cascade may occur not because a predator consumes a prey item, but because the prey species reduces foraging time to minimize risk, which results in a population increase at a lower trophic level. If the intermediate consumer or grazer elects not to forage in response to a top-predator, there will still be an increase in primary producers despite top-predators not actually consuming grazers.

Examples of non-lethal predator effects abound, and these effects have been shown to cause trophic cascades in aquatic and terrestrial communities. In oyster reefs, juvenile oysters (basal trophic level) are consumed by mud crabs (intermediate consumers), but predation on juvenile oysters is alleviated when toad fish (top-predators) are present. Toad fish consume mud crabs (lethal effect) and also cause mud crabs to seek refuge within the reef matrix and stop foraging (non-lethal effect). Both of these effects benefit juvenile oysters by reducing predation on them by mud crabs. In the fields of New England, spiders reduce grasshopper consumption of vegetation by eating grasshoppers, thereby reducing their numbers directly, and by causing the grasshoppers to seek refuge and stop foraging. Indeed the effects of wolves on elk grazing in Yellowstone Park appear to be mediated more by a reluctance of elk to venture into open meadows to forage than by direct predation on elk by wolves. Finally, flies in the family Phoridae are parasitoids of fire ants and many studies have examined their usefulness as biological control of fire ants. These flies decapitate fire ants, but, they also cause fire ant colonies cease foraging and individual ants remain in the nests when these flies are present, which reduces the ants' impacts in nature.

In these examples, it is clear that predators can have significant effects on the composition of entire communities by consuming lower trophic levels, and by altering the behavior or habitat selection of prey. Understanding how predators affect communities remains a central goal of contemporary ecology as changes in predator population densities or predator behavior may have significant effects on entire ecosystems. Many predator species are in decline globally, and conservation of these important species will likely be essential to insure the long-term stability of freshwater, marine, and terrestrial ecosystems.（Smee, 2010）

Glossary

dominant species 优势种
trophic 营养的
intertidal 潮间的
mussel 贻贝，蚌类
barnacle 藤壶
seaweed 海草，海藻
starfish 海星
biodiversity 生物多样性
kelp 海带
sea urchin 海胆
sea otter 海獭
food chain 食物链
trophic cascade 营养级联
Yellowstone National Park 黄石国家公园

primary producer 初级生产者
planktivorous fish 食浮游生物的鱼类
zooplankton 浮游动物
phytoplankton 浮游植物
deprive 使丧失，剥夺
overfish 过度捕捞
habitat 生境
biological community 生物群落
lethal effect 致命效应，杀伤效果
non-lethal effect 非致死效应
community composition 群落组成
oyster reef 牡蛎礁
Phoridae 蚤蝇科
biological control 生物防治

Unit 3 Community ecology

Lesson 10 Effects of biogeography on community diversity

Some places contain more species than others. For example, Antarctica has fewer species than a temperate deciduous forest, which in turn has fewer species a tropical rainforest. For over 150 years, researchers have sought to make sense of the gross and fine scale spatial patterns in biodiversity, and to elucidate both the proximate and ultimate causes for these patterns.

Many of the spatial patterns in biodiversity are overt, others are subtle and yet additional patterns remain undetected. While the existence of these patterns may be obvious — and changes in the environment that are paired with these patterns may also be obvious — the mechanisms that cause the differences in biodiversity along environmental gradients are under still the subject of scientific debate. Because large-scale patterns are the emergent result of complex interactions at many spatial and temporal scales, no single answer is likely to ever emerge, but with continued research our understanding of the processes shaping these patterns increases.

Major spatial patterns in biodiversity

One major geographic pattern in biodiversity is the latitudinal gradient in species richness. As one travels further away from the equator, for most taxa, the number of species declines. This general pattern holds true for most taxa and ecosystem types in both marine and terrestrial environments. There is broad agreement that this pattern is caused by differences in the abiotic, climatic environment, but the specific mechanism or mechanisms causing this pattern are a continued topic of discussion and investigation. One set of theories, broadly grouped together as "species-energy theory" is based on the fact that the amount of radiant energy from the sun captured by ecosystems is negatively associated with latitude. As energy is distributed throughout ecosystems through trophic processes, it is thought that species richness will track the energy following one or more mechanisms. Models of species-energy theory incorporate variables such as temperature, <u>net primary productivity</u>, <u>speciation</u>, and <u>extinction</u>. Other ideas that have been proposed to account for the latitudinal gradient are related to the physiological responses of animals to climatic conditions and the effects of the <u>abiotic environment</u> on historical processes. Most of these theories are not mutually exclusive.

Another recurring pattern in biogeographic theory is the <u>elevational gradient</u> in species richness. As one travels to higher elevations, the number of species declines, or, in many cases, peaks at mid-elevations. Aside from the environmental mechanisms driving this <u>diversity gradient</u>, there is a phenomenon that is based on the geography of species range distributions called the <u>mid-domain effect</u>. The mid-domain effect predicts a peak of diversity at the midpoint along any domain simply by the fact that the ranges of more species overlap in the middle of a domain (like a mountain or an island) than on the edges, and this effect works together with environmental determinants to affect the net distribution of species along many elevational gradients.

A third recurring pattern in the distribution of species is the area effect on species richness. The larger an place is, the more species it can support. This applies to actual islands in bodies of water, as well as habitat islands such as those surrounded by human development. The species-area relationship is presented in more detail in an article about <u>the theory of island biogeography</u>.

Do neutral models explain patterns in species richness?

So far, we have discussed biodiversity in terms of species richness but not whether the properties of the species in a location are related to the level of species richness. Are all species at a given trophic level interchangeable, in terms of their effects on overall species richness? A number of neutral models (such as the theory of island biogeography) do not consider the specific ecological interactions of members of a community. In other words, a <u>neutral theory</u> of biodiversity does not consider differences in the <u>niches</u> of any species at a given trophic position. Under neutral models, differences in relative abundance of any species are caused by historic patterns of abundance and dispersion, and not by the traits of any given species. Even if the niches of two species are demonstrably distinct, then under neutral models these species have equal effects on biodiversity. Neutral models may turn out to serve as effective null models for community assembly, just as Hardy-Weinberg models have served for population genetics. They also allow explicit modeling to test competing models for large-scale patterns in biodiversity.

McGlynn has also talked about large-scale effects on diversity — such as climate and area, but do local interactions — like competition and predation — influence the distribution of biodiversity? The fact that some species have greater effects on species richness than others may be gleaned from studying the biology of <u>invasive species</u>. An invasive species is one which is transported outside its original geographic range to a novel habitat, where it increases in density, and can cause detrimental effects on the indigenous species in that area, often reducing biodiversity. One example of an invasive species is the Argentine ant (*Linepithema humile*); more than 100 years ago this species was transported from Argentina to most other biogeographic regions, and where this species occur the local species of other ants is markedly reduced. Another example of an invasive species with negative effects on native faunas is cane toads (*Rhinella marina*) introduced from the Neotropics into Australia in 1935. Cane toads have reduced prey availability for indigenous predators and caused declines in native Australian amphibians. Most introduced species do not become invasive, indeed they frequently become extinct because they are less likely to be well adapted to the new environment than indigenous species, but the subset of introduced species that become invasive is problematic in terms of the economy, ecosystem services, and biodiversity. The widespread exchange of species among distant parts of the world as a result of human commerce has clearly resulted in a loss of biodiversity at many spatial scales.（McGlynn, 2010）

Glossary

biodiversity　生物多样性
spatial pattern　空间格局
environmental gradient　环境梯度
species richness　物种丰富度
taxon　分类，类群
net primary productivity　净初级生产力
speciation　物种形成
extinction　灭绝，衰减

abiotic environment　非生物环境
elevational gradient　海拔梯度
diversity gradient　多样性梯度
mid-domain effect　中域效应
the theory of island biogeography　岛屿生物地理学理论
neutral theory　中性理论
niche　生态位
invasive species　入侵种

Lesson 11　Ecological succession

Ecological succession is the process of change in the species structure of an ecological community over time. The time scale can be decades (for example, after a wildfire), or even millions of years after a mass extinction.

In <u>primary succession</u>, newly-exposed or newly-formed land is colonized by living things. In <u>secondary succession,</u> part of an ecosystem is disturbed, but remnants of the previous community remain.

Primary succession and pioneer species

Primary succession occurs when new land is formed or rock is exposed; for example, following the eruption of volcanoes, such as those on the Big Island of Hawaii. As lava flows into the ocean, new land is continually being formed. On the Big Island, approximately 32 acres of land are added each year. First, weathering and other natural forces break down the substrate enough for the establishment of certain hearty plants and lichens with few soil requirements, known as <u>pioneer species</u>. These species help to further break down the mineral-rich lava into soil where other, less-hardy species will grow, eventually replacing the pioneer species. In addition, as these early species grow and die, they add to an ever-growing layer of decomposing organic material, contributing to soil formation. Over time, the area will reach an equilibrium state with a set of organisms quite different from the pioneer species.

Secondary succession

Secondary succession is one of the two types of ecological succession of plant life. As opposed to the first, primary succession, secondary succession is a process started by an event (e.g. forest fire, harvesting, hurricane) that reduces an already established ecosystem (e.g. a forest or a wheat field) to a smaller population of species, and as such secondary succession occurs on preexisting soil whereas primary succession usually occurs in a place lacking soil.

A classic example of secondary succession occurs in oak and hickory forests cleared by wildfire. Wildfires will burn most <u>vegetation</u> and kill those animals unable to flee the area. Their nutrients, however, are returned to the ground in the form of ash. Thus, even when areas are devoid of life due to severe fires, they will soon be ready for new life to take hold.

Before a wildfire, vegetation is often dominated by tall trees with access to the major plant energy resource: sunlight. Their height gives them access to sunlight while also shading the ground and other low-lying species. After the fire, however, these trees are no longer dominant. Thus, the first plants to grow back are usually annual plants followed, within a few years, by quickly-growing and spreading grasses along with other pioneer species. Due to changes in the environment brought on by the growth of the grasses and other species, over many years, shrubs will emerge along with small pine, oak, and hickory trees. These organisms are called intermediate species. Eventually, over 150 years, the forest will reach its equilibrium point where species composition is no longer changing and resembles the community before the fire. This equilibrium state is referred to as the climax community, which will remain stable until the next disturbance.（https://www.boundless.com/biology/textbooks/boundless-biology-textbook/population-and-community-ecology-45/community-ecology-254/ecological-succession-939-12198/）

Glossary

ecological succession　生态演替　　　pioneer species　先锋物种
community　群落　　　　　　　　　　vegetation　植被
primary succession　原生演替　　　　　climax community　顶级群落
secondary succession　次生演替

Unit 4 Ecosystem ecology

Ecosystem ecology is the integrated study of living (biotic) and non-living (abiotic) components of ecosystems and their interactions within an ecosystem framework.

Ecosystem ecology examines physical and biological structures and examines how these ecosystem characteristics interact with each other. Ultimately, this helps us understand how to maintain high quality water and economically viable commodity production. A major focus of ecosystem ecology is on functional processes, ecological mechanisms that maintain the structure and services produced by ecosystems. These include primary productivity (production of biomass), decomposition, and trophic interactions.

Studies of ecosystem function have greatly improved human understanding of sustainable production of forage, fiber, fuel, and provision of water. Functional processes are mediated by regional-to-local level climate, disturbance, and management. Thus ecosystem ecology provides a powerful framework for identifying ecological mechanisms that interact with global environmental problems, especially global warming and degradation of surface water.

Lesson 12 Energy flow in ecosystems

Ecosystems maintain themselves by cycling energy and nutrients obtained from external sources. At the first trophic level, primary producers (plants, algae, and some bacteria) use solar energy to produce organic plant material through photosynthesis. Herbivores—animals that feed solely on plants—make up the second trophic level. Predators that eat herbivores comprise the third trophic level; if larger predators are present, they represent still higher trophic levels. Organisms that feed at several trophic levels (for example, grizzly bears that eat berries and salmon) are classified at the highest of the trophic levels at which they feed. Decomposers, which include bacteria, fungi, molds, worms, and insects, break down wastes and dead organisms and return nutrients to the soil.

On average about 10 percent of net energy production at one trophic level is passed on to the next level. Processes that reduce the energy transferred between trophic levels include respiration, growth and reproduction, defecation, and nonpredatory death (organisms that die but are not eaten by consumers). The nutritional quality of material that is consumed also influences how efficiently energy is transferred, because consumers can convert high-quality food sources into new living tissue more efficiently than low-quality food sources.

The low rate of energy transfer between trophic levels makes decomposers generally more important than producers in terms of energy flow. Decomposers process large amounts of organic material and return nutrients to the ecosystem in inorganic forms, which are then taken up again by primary producers. Energy is not recycled during decomposition, but rather is released, mostly as heat (this is what makes compost piles and fresh garden mulch warm).

An ecosystem's gross primary productivity (GPP) is the total amount of organic matter that it produces through photosynthesis. Net primary productivity (NPP) describes the amount of energy that remains available for plant growth after subtracting the fraction that plants use for respiration. Productivity in land ecosystems generally rises with temperature up to about 30℃, after which it declines, and is positively correlated with moisture. On

Unit 4　Ecosystem ecology

land primary productivity thus is highest in warm, wet zones in the tropics where tropical forest biomes are located. In contrast, desert scrub ecosystems have the lowest productivity because their climates are extremely hot and dry.

In the oceans, light and nutrients are important controlling factors for productivity. "Oceans" light penetrates only into the uppermost level of the oceans, so photosynthesis occurs in surface and near-surface waters. Marine primary productivity is high near coastlines and other areas where upwelling brings nutrients to the surface, promoting plankton blooms. Runoff from land is also a source of nutrients in estuaries and along the continental shelves. Among aquatic ecosystems, algal beds and coral reefs have the highest net primary production, while the lowest rates occur in the open due to a lack of nutrients in the illuminated surface layers.

How many trophic levels can an ecosystem support? The answer depends on several factors, including the amount of energy entering the ecosystem, energy loss between trophic levels, and the form, structure, and physiology of organisms at each level. At higher trophic levels, predators generally are physically larger and are able to utilize a fraction of the energy that was produced at the level beneath them, so they have to forage over increasingly large areas to meet their caloric needs.

Because of these energy losses, most terrestrial ecosystems have no more than five trophic levels, and marine ecosystems generally have no more than seven. This difference between terrestrial and marine ecosystems is likely due to differences in the fundamental characteristics of land and marine primary organisms. In marine ecosystems, microscopic phytoplankton carries out most of the photosynthesis that occurs, while plants do most of this work on land. Phytoplankton are small organisms with extremely simple structures, so most of their primary production is consumed and used for energy by grazing organisms that feed on them. In contrast, a large fraction of the biomass that land plants produce, such as roots, trunks, and branches, cannot be used by herbivores for food, so proportionately less of the energy fixed through primary production travels up the food chain.

Growth rates may also be a factor. Phytoplanktons are extremely small but grow very rapidly, so they support large populations of herbivores even though there may be fewer algae than herbivores at any given moment. In contrast, land plants may take years to reach maturity, so an average carbon atom spends a longer residence time at the primary producer level on land than it does in a marine ecosystem. In addition, locomotion costs are generally higher for terrestrial organisms compared to those in aquatic environments.

The simplest way to describe the flux of energy through ecosystems is as a food chain in which energy passes from one trophic level to the next, without factoring in more complex relationships between individual species. Some very simple ecosystems may consist of a food chain with only a few trophic levels. For example, the ecosystem of the remote wind-swept Taylor Valley in Antarctica consists mainly of bacteria and algae that are eaten by nematode worms. More commonly, however, producers and consumers are connected in intricate food webs with some consumers feeding at several trophic levels.

An important consequence of the loss of energy between trophic levels is that contaminants collect in animal tissues—a process called bioaccumulation. As contaminants bioaccumulate up the food web, organisms at higher trophic levels can be threatened even if the pollutant is introduced to the environment in very small quantities.

The insecticide DDT, which was widely used in the United States from the 1940s through the 1960s, is a famous case of bioaccumulation. DDT built up in eagles and other raptors to levels high enough to affect their reproduction, causing the birds to lay thin-shelled eggs that broke in their nests. Fortunately, populations have rebounded over several decades since the pesticide was banned in the United States. However, problems persist in some developing countries where toxic bioaccumulating pesticides are still used.

  Bioaccumulation can threaten humans as well as animals. For example, in the United States many federal and state agencies currently warn consumers to avoid or limit their consumption of large predatory fish that contain high levels of mercury, such as <u>shark</u>, <u>swordfish</u>, <u>tilefish</u>, and <u>king mackerel</u>, to avoid risking neurological damage and birth defect.
（http://learner.org/courses/envsci/unit/text.php?unit=4&secNum=3）

Glossary

trophic level　营养级
primary producer　初级生产者
photosynthesis　光合作用
defecation　排粪，排便
decomposer　分解者
producer　生产者
gross primary productivity　总初级生产力
net primary productivity　净初级生产力
aquatic ecosystem　水生生态系统

microscopic　微小的
phytoplankton　浮游植物
food chain　食物链
food web　食物网
contaminant　污染物
bioaccumulation　生物积累
insecticide　杀虫剂
shark　鲨鱼
swordfish　旗鱼
tilefish　方头鱼
king mackerel　鲭鱼

Lesson 13　Biogeochemical cycles

  Another major focus of ecosystem ecology is understanding how the chemical elements necessary for life persist and translocate in pools and fluxes within the ecosphere. The biosphere actively interacts with the three abiotic spheres (hydrosphere, atmosphere and lithosphere) to provide the available concentration of each for life. This action has a significant impact on the relative distribution of these elements. The simple sugar products of photosynthesis, $C_6H_{12}O_6$, are the base for organic matter, so carbon, hydrogen, and oxygen dominate the composition of life, and while oxygen is available in the lithosphere, and hydrogen in the hydrosphere, carbon is actually quite scarce in the environment, making the disproportionate amount of carbon in biomass a hallmark of life. In fact, there are about 20 elements used regularly in living organisms, of which nine called the macronutrients are the major constituents of organic matter: hydrogen, oxygen, carbon, nitrogen, calcium, potassium, silicon, magnesium, and phosphorus. Some of these elements are readily available in the abiotic environment, in which case conservation through cycling of the elements is not paramount; however, those in scarce supply, such as nitrogen and phosphorus, are reused many times before being released from the system. These biogeochemical cycles provide the foundation to understand how human modification leads to eutrophication (N and P cycles) and global climate change (C cycle). Therefore, much effort has been made to study and understand these cycles, particularly the carbon, nitrogen, and phosphorus cycles, details of which are addressed in here.

The Carbon Cycle

Carbon is an essential part of all organic molecules, and, as constituents of the atmosphere, carbon compounds such as carbon dioxide, CO_2, and methane, CH_4, substantially influence global climate. This connection between atmospheric carbon and climate has drawn all nations of the planet into discussions of the ecology of carbon cycling.

Carbon moves between organisms and the atmosphere as a

consequence of two reciprocal biological processes: photosynthesis and respiration. Photosynthesis removes CO_2 from the atmosphere, while respiration by primary producers and consumers, including decomposers, returns carbon to the atmosphere in the form of CO_2. In aquatic ecosystems, CO_2 must first dissolve in water before being used by aquatic primary producers. Once dissolved in water, CO_2 enters a chemical equilibrium with bicarbonate, HCO_3^-, and carbonate, CO_3^{2-}. Carbonate may precipitate out of solution as calcium carbonate and may be buried in ocean sediments.

While some carbon cycles rapidly between organisms and the atmosphere, some remains sequestered in relatively unavailable forms for long periods of time. Carbon in soils, peat, fossil fuels, and carbonate rock would generally take a long time to return to the atmosphere. During modern times, however, fossil fuels have become a major source of atmospheric CO_2 as humans have tapped into fossil fuel supplies to provide energy for their economic systems.

The nitrogen cycle

Nitrogen is important to the structure and functioning of organisms. It forms part of key biomolecules such as amino acids, nucleic acids, and the porphyrin rings of chlorophyll and hemoglobin. In addition, nitrogen supplies may limit rates of primary production in marine and terrestrial environments. Because of its importance and relative scarcity, nitrogen has drawn a great of attention from ecosystem ecologists.

Like the carbon cycle, the nitrogen cycle also includes a major atmospheric pool in the form of molecular nitrogen, N_2. However, only a few organisms can use this atmospheric supply of molecular nitrogen directly. These organisms, called nitrogen fixers, include (1) the cyanobacteria, or blue-green algae, of freshwater, marine, and soil environments, (2) free-living soil bacteria, (3) bacteria associated with the roots of leguminous plants, and (4) actinomycetes, filamentous bacteria, associated with the roots of alders, *Alnus*, and several other species of woody plants.

Because of the strong triple bonds between the two nitrogen atoms in the N_2 molecule, nitrogen fixation is an energy demanding process. During nitrogen fixation, N_2 is reduced to ammonia, NH_3. Nitrogen fixation takes place under aerobic conditions in terrestrial and aquatic environments, where nitrogen-fixing species oxidize sugars to obtain the required energy. Nitrogen fixation also occurs as a physical process associated with the high pressures and energy generated by lightning. Ecologists propose that all of the nitrogen cycling within ecosystems ultimately entered these cycles through nitrogen fixation by organisms or lightning. There is a relatively large pool of nitrogen cycled in the biosphere but only a small entryway through nitrogen fixation.

Once nitrogen is fixed by nitrogen-fixing organisms, it becomes available to other organisms within an ecosystem. Upon the death of an organism, the nitrogen in its tissues can be released by fungi and bacteria involved in the decomposition process. These fungi and bacteria release nitrogen as ammonium, NH_4^+, which may be converted to nitrate, NO_3^- by other bacteria. Ammonium and nitrate can be used directly by mycorrhizal fungi. The nitrogen absorbed by mycorrhizal fungi can be passed on to plants. The nitrogen in bacterial, fungal, and plant biomass may pass on to populations of animal consumers or back to the pool of dead organic matter, where it will be recycled again.

Nitrogen may exit the organic matter pool of an ecosystem through denitrification. Denitrification is an energy-yielding process that occurs under anaerobic conditions and converts nitrate to molecular nitrogen, N_2. The molecular nitrogen produced by denitrifying bacteria moves into the atmosphere and can only reenter the organic matter pool through nitrogen fixation. Ecologists estimate that the mean residence time of fixed nitrogen in the biosphere is about 625 years. They estimate that the mean residence time of phosphorus in the biosphere is on the order of thousands of years.

The phosphorus cycle

Phosphorus is essential to the energetics, genetics, and structure of living systems. For instance, phosphorus forms part of the ATP, RNA, DNA, and phospholipid molecules. While of great biological importance, phosphorus is not very abundant in the biosphere. Consequently, phosphorus cycling has received a great deal of attention from ecosystem ecologists.

In contrast to carbon and nitrogen, the global phosphorus cycle does not include a substantial atmospheric pool. The largest quantities of phosphorus occur in mineral deposits and marine sediments. Sedimentary rocks that are especially rich in phosphorus are mined for fertilizer and applied to agricultural soils. Soil may contain substantial quantities of phosphorus. However, much of the phosphorus in soils occurs in chemical forms not directly available to plants.

Phosphorus is slowly released to terrestrial and aquatic ecosystems through the weathering of rocks. As phosphorus is released from mineral deposits, it is absorbed by plants and recycled within ecosystems. However, much phosphorus is washed into rivers and eventually finds its way to the oceans, where it will remain in dissolved form until eventually finding its way to the ocean sediments. Ocean sediments will be eventually transformed into phosphate-bearing sedimentary rocks that through geological uplift can form new land. William Schlesinger points out that the phosphorus released by the weathering of sedimentary rocks has made at least one passage through the global phosphorus cycle. (Jørgensen SE., 2011; Molles M., 2000)

Glossary

ecosphere 生态圈，生物圈
macronutrient 大量元素
biogeochemical cycle 生物地球化学循环
eutrophication 富营养化
biomolecule 生物分子
amino acid 氨基酸
nucleic acid 核酸
porphyrin ring 卟啉环

chlorophyll 叶绿素
hemoglobin 血红蛋白
cyanobacteria 蓝藻
leguminous 豆科的
actinomycete 放线菌
mycorrhizal fungi 菌根真菌
denitrification 反硝化作用
weathering 风化

Lesson 14　Ecosystem dynamics

Ecosystems are dynamic entities—invariably, they are subject to periodic <u>disturbances</u> and are in the process of recovering from some past disturbance. When an ecosystem is subject to some sort of <u>perturbation</u>, it responds by moving away from its initial state. The tendency of a system to remain close to its <u>equilibrium</u> state, despite that disturbance, is termed its resistance. On the other hand, the speed with which it returns to its initial state after disturbance is called its <u>resilience</u>.

From one year to another, ecosystems experience variation in their biotic and abiotic environments. A drought, an especially cold winter and a pest outbreak all constitute short-term variability in environmental conditions. Animal populations vary from year to year, building up during resource-rich periods and crashing as they overshoot their food supply. These changes play out in changes in net primary productivity (NPP), decomposition rates, and other ecosystem processes. Longer-term changes also shape ecosystem processes—the forests of eastern North America still show legacies of cultivation which ceased 200 years ago, while methane production in eastern Siberian lakes is controlled by organic matter which accumulated during the Pleistocene.

Disturbance also plays an important role in ecological processes. Chapin and coauthors define disturbance as "a relatively discrete event in time and space that alters the structure of populations, communities and ecosystems and causes changes in resources availability or the physical environment". This can range from tree falls and insect outbreaks to hurricanes and wildfires to volcanic eruptions and can cause large changes in plant, animal and microbe populations, as well soil organic matter content. Disturbance is followed by succession, a "directional change in ecosystem structure and functioning resulting from biotically driven changes in resources supply".

The frequency and severity of disturbance determines the way it impacts ecosystem function. Major disturbances like a volcanic eruption or glacial advance and retreat leave behind soils that lack plants, animals or organic matter. Ecosystems that experience disturbances undergo primary succession. Less severe disturbance like forest fires, hurricanes or cultivation result in secondary succession. More severe disturbance and more frequent disturbance result in longer recovery times. Ecosystems recover more quickly from less severe disturbance events.

The early stages of primary succession are dominated by species with small propagules (seed and spores) which can be dispersed long distances. The early colonizers—often algae, cyanobacteria and lichens—stabilize the substrate. Nitrogen supplies are limited in new soils, and nitrogen-fixing species tend to play an important role early in primary succession. Unlike in primary succession, the species that dominate secondary succession are usually present from the start of the process, often in the soil seed bank. In some systems the successional pathways are fairly consistent, and thus, are easy to predict. In others, there are many possible pathways—for example, the introduced nitrogen-fixing legume, *Myrica faya*, alters successional trajectories in Hawaiian forests.

The theoretical ecologist Ulanowicz has used information theory tools to describe the structure of ecosystems, emphasizing mutual information (correlations) in studied systems. Drawing on this methodology and prior observations of complex ecosystems, Ulanowicz depicts approaches to determining the stress levels on ecosystems and predicting system reactions to defined types of alteration in their settings (such as increased or reduced energy flow, and eutrophication).（https://en.wikipedia.org/wiki/ Ecosystem）

 Glossary

disturbance 干扰	succession 演替
perturbation 忧虑，不安	primary succession 初生演替
equilibrium 平衡	secondary succession 次生演替
resistance 阻力，抵抗	propagule 繁殖体
resilience 弹性，弹力，快速回复的能力	spore 孢子
	algae 水藻
Pleistocene 更新世	cyanobacteria 蓝藻
hurricane 飓风	lichen 地衣
wildfire 野火	seed bank 种子库
volcanic eruption 火山喷发	energy flow 能量流
microbe 微生物	eutrophication 富营养化

Lesson 15　Ecosystem services

Humans have always relied on nature for environmental assets like clean water and soil formation. Today, these assets are receiving global attention as "ecosystem services", the conditions and processes by which natural ecosystems sustain and fulfill human life. Natural ecosystems perform a diversity of ecosystem services on which human civilization depends:

1. regulating services—purification of air and water, detoxification and decomposition of wastes, moderation of weather extremes, climate regulation, erosion control, flood control, mitigation of drought and floods, regulation of disease carrying organisms and agricultural pests;

2. provisioning services—provision of food, fuel, fiber, and freshwater;

3. supporting services—formation and preservation of soils, protection from ultraviolet rays, pollination of natural vegetation and agricultural crops, cycling of nutrients, seed dispersal, maintenance of biodiversity, primary production;

4. cultural services—spiritual, esthetic, recreational.

Based on the estimate that one-third of human food comes from plants pollinated by wild pollinators, pollination has been valued at US4-6 billion per year in the US alone. Globally, the world's ecosystem services have been valued at US33 trillion a year, nearly twice as much as the gross national product of all of the world's countries.

The idea of paying for ecosystem services has been gaining momentum. Yet, because ecosystem services are typically not sold in markets, they usually lack a market value. Given the value of natural capital, nonmarket valuation approaches are being developed by economists and ecologists to account for ecosystem services in decision-making processes. The notion being that economic valuation gives decision makers a common currency to assess the relative importance of ecosystem processes and other forms of capital.

Yet, assigning value to ecosystem services is tricky and some analysts object to nonmarket valuation, because it is a strictly anthropogenic measure and does not account for nonhuman values and needs. Yet, in democratic countries, environmental policy outcomes are determined by the desires of the majority of citizens, and voting on a preferred policy alternative is ultimately an anthropogenic activity. A second objection to nonmarket valuation is a disagreement with pricing the natural world and dissatisfaction with the capitalistic premise that everything is thought of in terms of commodities and money. The point of valuation, however, is to frame choices and clarify the tradeoffs between alternative outcomes. Finally, a third objection to nonmarket valuation stems from the uncertainty in identifying and quantifying all ecosystem services. Advocates argue that economic valuation need not cover all values and that progress is made by capturing values that are presently overlooked.

Ecosystem services are threatened by growth in the scale of human enterprise (population size, per-capita consumption rates) and a mismatch between short-term needs and long-term societal well-being. With a global population soon to number 9 billion people, ecosystem services are becoming so degraded, some regions in the world risk ecological collapse. Many human activities alter, disrupt, impair, or reengineer ecosystem services such as overfishing, deforestation, introduction of invasive species, destruction of wetlands, erosion of soils, runoff of pesticides, fertilizers, and animal wastes, pollution of land, water, and air resources. The consequences of degrading ecosystem services on human well-being were examined in the Millennium Ecosystem Assessment (MA) 2005, which concluded that well over half of the world's ecosystems services are being degraded or used unsustainably. The MA developed global ecological scenarios as a process to inform future policy options. These scenarios were based on a suite of models that were designed to forecast future change. The MA based its scenario analyses on ecosystem services. Specifically, scenarios were developed to anticipate responses of ecosystem services to alternative futures driven by different sets of policy decisions. Following the completion of this ambitious ecological study, there is now a growing movement to make the value of ecosystem services an integral pan of current policy initiatives . (Jørgensen, 2011)

Glossary

ecosystem service 生态系统服务	产生的
civilization 文明，文化	democratic 民主的，民主党的
detoxification 消毒	capitalistic 资本主义的
erosion 侵蚀	enterprise 企业，事业
pest 害虫	overfishing 过渡捕捞
freshwater 淡水	deforestation 采伐森林
pollinator 传粉者，传粉媒介	wetland 湿地
anthropogenic 人为的，人类活动	

Lesson 16 Evolution and natural selection in ecosystems

As species interact, their relationships with competitors, predators, and prey contribute to natural selection and thus influence their evolution over many generations. To illustrate this concept, consider how <u>evolution</u> has influenced the factors that affect the foraging efficiency of predators. This includes the predator's search time (how long it takes to find prey), its handling time (how hard it has to work to catch and kill it), and its prey profitability (the ratio of energy gained to energy spent handling prey).

Characteristics that help predators find, catch, and kill prey will enhance their chances of surviving and reproducing. Similarly, prey will profit from attributes that help avoid detection and make organisms harder to handle or less biologically profitable to eat.

These common goals drive natural selection for a wide range of traits and behaviors, including: (1) Mimicry by either predators or prey. A predator such as a praying mantis that blends in with surrounding plants is better able to surprise its target. However, many prey species also engage in mimicry, developing markings similar to those of unpalatable species so that predators avoid them. For example, harmless viceroy butterflies have similar coloration to monarch butterflies, which store toxins in their tissues, so predators avoid viceroy butterflies. (2) Optimal foraging strategies enable predators to obtain a maximum amount of net energy per unit of time spent foraging. Predators are more likely to survive and reproduce if they restrict their diets to prey that provide the most energy per unit of handling time and focus on areas that are rich with prey or that are close together. The Ideal Free Distribution model suggests that organisms that are able to move will distribute themselves according to the amount of food available, with higher concentrations of organisms located near higher concentrations of food. Many exceptions have been documented, but this theory is a good general predictor of animal behavior. (3) Avoidance/escape features help prey elude predators. These attributes may be behavioral patterns, such as animal herding or fish schooling to make individual organisms harder to pick out. Markings can confuse and disorient predators: for example, the automeris moth has false eye spots on its hind wings that misdirect predators. (4) Features that increase handling time help to discourage predators. Spines serve this function for many plants and animals, and shells make crustaceans and mollusks harder to eat. Behaviors can also make prey harder to handle: squid and octopus emit clouds of ink that distract and confuse attackers, while <u>hedgehogs</u> and <u>porcupines</u> increase the effectiveness of their protective spines by rolling up in a ball to conceal their vulnerable underbellies. (5) Some plants and animals emit noxious chemical substances to make themselves less profitable as prey. These protective substances may be bad-tasting, <u>antimicrobial</u>, or <u>toxic</u>. Many species that use noxious substances as protection have evolved bright coloration that signals their identity to would-be predators—for example, the black and yellow coloration of bees, <u>wasps</u>, and yellow jackets. The substances may be generalist defenses that protect against a range of threats, or specialist compounds developed to ward off one major predator. Sometimes specialized predators are able overcome these noxious substances: for example, ragwort contains toxins that can poison horses and cattle grazing on it, but it is the exclusive food of cinnabar moth caterpillars. Ragwort toxin is stored in the caterpillars' bodies and eventually protects them as moths from being eaten by birds.

Natural selection based on features that make predators and prey more likely to survive can generate predator-prey "arms races" with improvements in prey defenses triggering counter-improvements in predator attack tools and vice versa over many generations. Many cases of predator-prey arms races have been identified. One widely known case is <u>bats</u>' use of <u>echolocation</u> to find insects. Tiger moths respond by emitting high-frequency clicks to jam bats' signals, but some bat species have overcome these measures through new techniques such as flying erratically to confuse moths or sending echolocation chirps at frequencies that moths cannot detect. This type of pattern involving two species that interact in important ways and evolve in a series of reciprocal genetic steps is called <u>coevolution</u> and represents an important factor in adaptation and the evolution of new biological species.

Other types of relationship, such as competition, also affect evolution and the characteristics of individual species. For example, if a species has an opportunity to move into a vacant <u>niche</u>, the shift may facilitate evolutionary changes over succeeding generations because the species plays a different ecological role in the new niche. By the early 20th century, large predators such as wolves and puma had been largely eliminated from the eastern United States. This has allowed coyotes, who compete with wolves where they are found together, to spread throughout urban, suburban, and rural habitats in the eastern states, including surprising locations such as Cape Cod in Massachusetts and Central Park in New York City. Research suggests that northeastern coyotes are slightly larger than their counterparts in western states, although it is not yet clear whether this is because the northeastern animals are hybridizing with wolves and domestic dogs or because they have adapted genetically to preying on larger species such as white-tailed deer.（http://learner.org/courses/envsci/unit/text.php?unit=4&secNum=8）

Glossary

evolution 演变，进化
behavior 行为
mimicry 模仿
butterfly 蝴蝶
crustacean 甲壳纲动物
mollusk 软体动物类
squid 乌贼
octopus 章鱼
hedgehog 刺猬

porcupine 豪猪
antimicrobial 抗菌的
toxic 有毒的
wasp 黄蜂
bat 蝙蝠
echolocation 回声
coevolution 协同进化
niche 生态位

Unit 5 Global change ecology

Humans are altering biological diversity, land cover, atmospheric composition, and the climate system at an unprecedented rate. Global change ecology is the study of how these alterations influence the complex web of interactions among species, ecosystem processes and the Earth as a whole. It is critical for us to understand these interactions so that science can provide a foundation for environmental policy at local, national, and international levels. Losses of biodiversity and changes in ecosystem goods and services represent some of the most pressing issues that humans face over the next several centuries.

Lesson 17 Global change: an overview

We live in a world where humans are having profound impacts on the global environment. Climate is warming, the populations of many species are in decline, pollution is affecting ecosystems and human health, and human societies now face new risks in terms of sea level changes, disease, food security, and climate extremes.

Scientists who study global environmental change are interested in learning how drivers of environmental change impact biological systems across many scales—from the level of the individual organism, to populations, communities, and ecosystems. Global environmental change science is therefore a highly <u>multidisciplinary</u> effort, involving physical scientists who study climate, the oceans, the atmosphere, and geology, as well as biologists investigating physiology, evolution, and ecology.

Drivers of global change

Human population and consumption

Almost 7 billion people now live on Earth. Rapid growth of the human population, especially over the last 300 years, is one of the most remarkable trends in population change ever observed. Demographers project that world population will rise to 9 billion by 2050 and level off somewhere between 9–12 billion people by the end of the century. In many modern societies, more people require more resources, such as crops, seafood, forest products, energy, and minerals and increasingly larger economies to support economic development and rising standards of living. Population growth and the increased demand for natural resources is therefore a major factor driving global environmental change.

The population story is more complex, however, because there is not a simple relationship between the number of people and the amount of resources consumed. Affluence, or the wealth per person, and the social norms of consumption are also important. For example, the populations of China and India are roughly 1.32 and 1.14 billion people, respectively—about four times that of the US. However, the energy consumption per person in the US is six times larger than that of a person in China, and 15 times that of a person in India. Because the demand for resources like energy is often greater in wealthy, developed nations like the US, this means that countries with smaller populations can actually have a greater overall environmental impact. Over much of the past century, the US was the largest greenhouse gas emitter because of high levels of affluence and energy consumption. In 2007, China overtook the US in terms of overall CO_2 emissions as a result of economic development, increasing personal wealth, and the demand for consumer goods, including automobiles.

Energy use/Climate change

Worldwide, <u>fossil fuels</u> (oil, coal, and natural gas) dominate our energy consumption, accounting for 85% of all energy used. As mentioned previously, the rapid rise of fossil fuels is a relatively recent phenomenon, developing in the nineteenth century with the discovery of oil and the industrialization of economies, and expanding rapidly in the twentieth century with increased economic development and rising populations and affluence. From 1860 to 1991, energy use per person rose more than 93 fold compared to a world population increase of four fold, indicating that rising affluence and consumption are driving energy demand.

Burning fossil fuels releases about 8.5 billion tons of carbon (as CO_2) into the atmosphere each year, causing its concentration to increase and Earth's greenhouse warming to strengthen, which leads to rising global air temperatures. Since 1880, average global air temperature has risen approximately 0.9℃. The top five CO_2-emitting countries/regions are China, US, EU, Russia, and India, which together account for two thirds of global emissions.

Global change scientists use climate models to determine how added greenhouse gases affect changes in air temperature and precipitation. If fossil fuel burning continues at current rates, global temperatures may rise by as much as 4℃ by the year 2100 (IPCC 2007). Precipitation changes are expected to lead to increased rainfall in mid-to-high latitude regions, but increased droughts are projected for subtropical regions (IPCC 2007).

Land use changes

Landscapes are changing worldwide, as natural land covers like forests, grasslands, and deserts are being converted to human-dominated ecosystems, including cities, agriculture, and forestry. Between 2000–2010, approximately 13 million hectares of land were converted each year to other land cover types (FAO 2010). Developed regions like the US and Europe experienced significant losses of forest and grassland cover over the past few centuries during phases of economic growth and expansion. More recently, developing nations have experienced similar losses over the past 60 years, with significant forest losses in biologically diverse regions like Southeast Asia, South America, and Western Africa.

Land use changes affect the <u>biosphere</u> in several ways. They often reduce native <u>habitat</u>, making it increasingly difficult for species to survive. Some land use changes, such as <u>deforestation</u> and agriculture, remove native vegetation and diminish carbon uptake by <u>photosynthesis</u> aswell as hasten soil decomposition, leading to additional greenhouse gas release. Almost 20% of the global CO_2 released to the atmosphere (1.5–2 billion tons of carbon) is thought to come from deforestation.

Pollution

One of the <u>byproducts</u> of economic development has been the production of pollution—products and waste materials that are harmful to human and ecological health. The rise of pollution corresponds to the increased use of petroleum in the twentieth century, as new synthetic products such as plastics, <u>pesticides</u>, solvents, and other chemicals, were developed and became central to our lives. Many air pollutants, including nitrogen and sulfur oxides, fine particulates, lead, carbon monoxide, and ground-level ozone come from coal and oil consumption by power plants and automobiles. <u>Heavy metals</u>, such as mercury, lead, cadmium, and arsenic, are produced from <u>mining</u>, the burning of fossil fuels, and the manufacture of certain products like metals, paints, and batteries.

<u>Aquatic ecosystems</u> such as rivers, lakes, and coastal oceans have

Unit 5 Global change ecology

traditionally been used for pollution disposal from industry and sewage treatment plants, but they have also been subject to unintentional runoff from upland watersheds, such as nitrogen and phosphorus loss from agricultural soils and home septic systems as well as plastics washed into rivers and oceans from storm sewer systems. We often don't think of nutrients like nitrogen and phosphorus as pollutants. However, humans now add more nitrogen to the biosphere through fertilizers than is added naturally each year by all of the <u>nitrogen-fixing bacteria</u> on the planet. The Pacific and Atlantic oceans now have garbage patches full of plastic that are possibly as large as the continental US. These are strong indicators of global change — humanity now dominates the global movement of nitrogen and other materials on Earth.（Camill, 2010）

Glossary

multidisciplinary　多学科，多部门
demographer　人口统计学家
greenhouse gas　温室气体
fossil fuel　化石燃料
biosphere　生物圈
habitat　生境
deforestation　采伐森林

photosynthesis　光合作用
byproduct　副产品
pesticide　杀虫剂
heavy metal　重金属
mining　采矿
aquatic ecosystem　水生生态系统
nitrogen-fixing bacteria　固氮细菌

Lesson 18　Human impact and ecosystem degradation

The monitoring of ecosystems and research on their functioning can be regarded as a comprehensive type of biological study that is designed to identify the interactions between the biotic and abiotic elements within a natural environment, and to trace the temporal and spatial changes in these elements. The task of monitoring the gradual changes in the environmental elements present in the atmosphere, water, and soil occasioned by changes in the global environment has yielded numerous ecological results.

Human activities have had the effect of bringing about rapid changes in the global environment. Human activities have unavoidably caused damage to ecosystems and the destruction of habitats, for example through such means as land development projects. The negative influence of such human activities has led in many cases to continuous changes in ecosystems. Amongst such negative influential factors can be included the decrease of biological populations as a result of excessive hunting and gathering, changes in forest use caused by development, destruction of ecosystems as a result of land reclamation, degradation of ecosystems due to the emission of environmental pollutants, and the decrease, erosion, and destruction of biological habitats because of the fragmentation and isolation of habitat patches.

Rapid urbanization has created a situation in which the transformation of the manner in which land in surrounding areas is used has continued unabated. For instance, land-reclamation projects have inflicted seemingly endless damage upon a major component of Korea's coastal ecosystem, namely its tidal and mud-flat areas. Semi-natural forests and many secondary forests possess high biodiversity; however, their fragmentation by roads and the development of residential areas has often severed the continuity of forests and watersheds, and led to the overall degradation of many of these terrestrial ecosystems. In addition, environmental pollutants have helped to degrade many ecosystems, and the emergence of global warming through greenhouse gas emissions has created potentially the major threat to ecosystems worldwide.

Unit 5　Global change ecology

Estuaries and many freshwater wetlands have been markedly degraded by human activities. For example, these types of ecosystems have been directly affected by land-reclamation projects. In addition, water pollution and soil inflow due to increased land use have also affected the estuary and tidal flat ecosystems. Imprudent human activities and interference have led to the degradation of ecosystems and habitats in primeval forests, even in remote high mountain ecosystems such as those found in the Himalayas. In addition to the existence of a direct causation of the loss of biodiversity, human activities at the landscape level have also caused changes in major biogeochemical cycles, for example in the carbon and water cycles. Such changes can have major impacts on the biota and on ecological processes in the long term, and there is the urgent need to minimize and if possible reverse these changes, in some cases through the restoration of degraded ecosystems.（Hong and Lee, 2006）

 **Glossary**

biotic　生物的	estuary　（江河入海的）河口，河口湾
abiotic　非生物的	
global environment　全球环境	freshwater wetland　淡水湿地
land reclamation　土地复垦	primeval forest　原始森林
fragmentation　分裂，破碎	Himalayas　喜马拉雅山脉
habitat patch　生境斑块	biogeochemical cycle　生物地球化学循环
urbanization　城市化	
coastal ecosystem　海岸带生态系统	biota　生物群
biodiversity　生物多样性	degraded ecosystem　退化生态系统
global warming　全球变暖	

Lesson 19 Impacts of global change on human society

Global change scientists are interested in understanding several potential impacts of climate warming on human societies, including human health, natural disturbances, and food security.

Health

Climate change and pollution pose a number of potential risks. One of the most commonly cited examples is the potential spread and rise of infectious diseases, such as the mosquito-borne tropical diseases, malaria and dengue fever. Warming temperatures are expected to increase the geographic range suitable for the mosquito species that transmit these diseases, and some scientists have warned that this could mean a spread of tropical diseases into areas that traditionally do not experience them, such as North America. However, other scientists have argued that the range of malaria actually contracted during twentieth-century warming. The reason, they argue, is that control measures, such as insecticides, bed nets, and medical treatments, are being deployed effectively to combat the diseases. Exposure to extreme heat is another risk to societies. The heat waves of 1995 in Chicago and 2003 in France caused 700 and 14,800 deaths, respectively. Poor sanitation increases the risk of cholera and other diarrheal diseases where floods contaminate water supplies with untreated sewage. The risk of these diseases rises with the increased occurrence of extreme precipitation events. Other causes of concern include increased respiratory illnesses resulting from ground-level ozone (smog) and fine particulates from vehicle exhaust and indoor cooking fires in urban and suburban areas.

Unit 5　Global change ecology

Natural disturbances

Changing climate has the potential to increase risks from <u>sea level rise</u>, extreme storm events, and <u>drought</u>. About 25% of the world's population lives with 100 km of the coast. In 2007, the IPCC projected an 18–59 cm sea level rise by 2100, but many scientists argue that this range is too low. People living in low lying regions, such as Bangladesh and Pacific island nations (e.g., Tuvalu and the Maldives) are already experiencing the effects of salt water incursion in their agricultural fields and fresh water supplies. Arctic Inuit communities are battling the loss of coastal villages as a result of increased <u>storm surges</u> from sea level rise.

Extreme precipitation events may become more common in mid-to-high latitude regions, consistent with the prediction that warmer air temperatures, as a result of climate warming, will increase the moisture content of the atmosphere (IPCC 2007). Besides catastrophic flooding and associated property damage, extreme storms are a concern for infrastructure, with cities now faced with the prospect of significant costs associated with roads, dams, and levees being washed out by floods. Meanwhile, drought is gripping subtropical regions worldwide, and precipitation is expected to decline in these regions over the twenty-first century (IPCC 2007). The American Southwest has been experiencing drought conditions since 1999, a duration comparable to other severe historical droughts such as the <u>Dust Bowl</u> of the 1930s in the American Great Plains. Warming is also diminishing the snow packs of the Rocky and Sierra Mountains, further decreasing water supplies to major metropolitan areas. Australia has also experienced severe dryness over the past decade. Droughts in climatically sensitive subtropical regions, such as sub Saharan Africa pose risks for food security for a number of African nations.

Global change scientists have also investigated whether <u>tropical cyclones</u> (hurricanes) are becoming more frequent or severe with climate change. Although observational and modeling studies suggest that hurricanes may become stronger in the future as a result of warmer sea surface temperatures, there is little evidence to suggest that they are becoming more common.

Food security

Food security is the ability of people to have access to sufficient, nutritious food. Although we currently grow enough food to feed the global human population, a population rising to 9 billion by 2050, combined with climate changes, will strain the capacity of some regions to feed people, thereby raising the risks of food insecurity.

Water is very important to food security because it is one of the key determinants of crop yield, and changing climate has the potential to alter precipitation around the world. It takes approximately 1,000 tons of water to produce a kilogram of grain, and 2–7 kilograms of grain are required per kilogram of livestock weight gain, suggesting that food production, and meat production in particular, requires substantial inputs of water from precipitation. Where precipitation is insufficient, irrigation, and the diversion of rivers, lakes, or groundwater is required to supplement rainfall.

The world faces several challenges with respect to water and food security. Warming temperatures and decreased precipitation may cause crop yields to decline, particularly in southern Asia and Africa, leading to rising food prices and increased difficulty in gaining access to food. In some of the world's largest grain producing regions, such as the American Southern Great Plains and the North China Plain, groundwater aquifers are being depleted by as much as 10 feet (3m) per year as irrigation is outpacing recharge by precipitation. Collapse of these aquifers would impact globally significant grain supplies. Poorly designed irrigation systems can also destroy important aquatic ecosystems, such as the Aral

Unit 5　Global change ecology

Sea in Asia, which was once one of the largest lakes in the world but has shrunk to 10% of its original size due to water diversion for agricultural irrigation. Population growth and rising affluence in developing nations and the growing demand for meat further exacerbates water demands. （Camill, 2010）

Glossary

natural disturbance 自然干扰	sea level rise 海平面上升
food security 食品安全	drought 干旱
malaria 疟疾	storm surge 风暴潮
dengue fever 登革热	Dust Bowl 沙尘暴
insecticide 杀虫剂	tropical cyclone 热带气旋
cholera 霍乱	groundwater aquifer 地下水含水层
diarrheal 腹泻的	aquatic ecosystem 水生生态系统
ozone 臭氧	Aral Sea 咸海
fine particulate 细颗粒物	

Lesson 20　Biodiversity and climate change

Over the past decade, several models have been developed to predict the impact of climate change on biodiversity. Results from these models have suggested some alarming consequences of climate change for biodiversity, predicting, for example, that in the next century many plants and animals will go extinct and there could be a large-scale dieback of tropical rainforests. However, caution may be required in interpreting results from these models, not least because their coarse spatial scales fail to capture topography or "microclimatic buffering" and they often do not consider the full acclimation capacity of plants and animals. Several recent studies indicate that taking these factors into consideration can seriously alter the model predictions. In one study, Randin et al. assessed the influence of spatial scale on the accuracy of bioclimatic model predictions of habitat losses for alpine plant species in the Swiss Alps. A coarse European-scale model (with 16 km by 16 km grid cells) predicted a loss of all suitable habitats during the 21st century, whereas a model run using local-scale data (25 m by 25 m grid cells) predicted persistence of suitable habitats for up to 100% of plant species. The authors attributed these differences to the failure of the coarser spatial-scale model to capture local topographic diversity, as well as the complexity of spatial patterns in climate driven by topography.

Luoto and Heikkinen reached a similar conclusion in their study of the predictive accuracy of bioclimatic envelope models (which model the relation between current climate variables and present-day species distributions) on the future distribution of 100 European butterfly species. A model that included climate and topographical heterogeneity (such as elevational range) predicted only half of the species losses in mountainous areas for the period from 2051 to 2080 in comparison to a climate-only model. In contrast, the number of species predicted to disappear from flatlands doubled in the climate-topography model relative to the climate-only model. The two studies suggest that habitat heterogeneity resulting from topographic diversity may be essential for persistence of biota in a future changing climate.

Highly contrasting predictions have also been obtained when bioclimatic models of tropical biomes included the physiological effects of elevated atmospheric CO_2 concentrations and temperature on trees. Many studies have indicated that increased atmospheric CO_2 affects photosynthesis rates and enhances net primary productivity—more so in tropical than in temperate regions—yet previous climate-vegetation simulations did not take this into account.

To address these issues, Lapola et al. developed a new potential-vegetation model for tropical South America that includes CO_2 fertilization effects. They then drove this model with different climate scenarios for the end of the 21st century from 14 coupled ocean-atmosphere global climate models of the Intergovernmental Panel on Climate Change (IPCC) Fourth Assessment report. The results indicate that when the CO_2 fertilization effects are considered, they overwhelm the impacts arising from temperature; rather than the large-scale die-back predicted previously, tropical rainforest biomes remain the same or substituted by wetter and more productive biomes. However, for 2 of the 14 models, this result was dependent on the dry season not extending beyond 4 months; if it does, then the tropical biome becomes savanna.

These studies highlight the level of complexity that we are faced with in trying to model and predict the possible consequences of future climate change on biodiversity. They suggest that we should expect to see species turnover, migrations, and novel communities, but not necessarily the levels of extinction previously predicted. For example, Hole et al. recently studied model-projected shifts in the distribution of sub-Saharan Africa's breeding bird fauna. They found that in the Important Bird Area protected network, species turnover is likely to be substantial and regionally variable, but persistence of suitable climate space across the network as a whole is remarkably high, with 88 to 92% of species retaining suitable climate space.

Another complexity, however, is the impact of climate change on already highly altered fragmented landscapes outside of protected areas. Over 75% of the Earth's terrestrial biomes now show evidence of alteration as a result of human residence and land use. Yet, recent case studies suggest that even in a highly fragmented landscape, all is not lost for biodiversity.

It has long been assumed that in a fragmented landscape, the fragment size and its isolation are important factors in determining species persistence; the smaller and more isolated the fragment, the lower its occupancy. Yet few worldwide studies have attempted to quantify this relation. Prugh et al. compiled and analyzed raw data from previous research on the occurrence of 785 animal species in >12,000 discrete habitat fragments on six continents. In many cases, fragment size and isolation were poor predictors of occupancy. The quality of the matrix surrounding the fragment had a greater influence on persistence: When the matrix provided conditions suitable to live and reproduce, fragment size and isolation were less important and species were able to persist.

This ability of species to persist in what would appear to be a highly undesirable and fragmented landscape has also been recently demonstrated in West Africa. In a census on the presence of 972 forest butterflies over the past 16 years, Larsen found that despite an 87% reduction in forest cover, 97% of all species ever recorded in the area are still present. For reasons that are not entirely clear, these butterfly species appear to be able to survive in the remaining primary and secondary forest fragments and disturbed lands in the West African rainforest. However, presence or absence does not take into account lag effects of declining populations; a more worrying interpretation is therefore that the full effects of fragmentation will only be seen in future years.

Predicting the fate of biodiversity in response to climate change combined with habitat fragmentation is a serious undertaking fraught with caveats and complexities. The recent studies discussed here attempt to quantify some of the uncertainty in these predictions. They use larger, more detailed data sets and more refined models than previously available, thus avoiding the problems often encountered in trying to scale up results from small local-scale studies.

The results also highlight a serious issue for future conservationists: the urgent need to develop a research agenda for regions outside of protected reserves in human-modified landscapes. Although every measure should be put in place to reduce further fragmentation of reserves, we must determine what represents a "good" intervening matrix in these human-modified landscapes. Furthermore, with the combination of climate change and <u>habitat destruction</u>, novel ecosystems are going to become increasingly common. Their conservation will require a whole new definition of what is "natural".（Willis and Bhagwat, 2009）

Glossary

biodiversity　生物多样性
extinct　灭绝的，绝种的
topography　地形，地貌
spatial scale　空间尺度
alpine plant　高山植物
spatial pattern　空间格局
heterogeneity　异质性，不均匀性
biota　生物群

photosynthesis　光合作用
net primary productivity　净初级生产力
migration　迁移
fauna　动物区系
habitat fragment　生境片段
rainforest　雨林
habitat destruction　生境破坏

Lesson 21　Global climate change and risk assessment: invasive species

<u>Invasive species</u> are considered the second greatest agent of change to ecosystems after <u>habitat</u> change. They can have both direct and indirect effects, resulting in ecosystem impacts defined as substantial impacts to

species composition, relative abundances, nutrient pools and fluxes, and disturbance cycles such as terrestrial fire regimes. They can also change contaminant cycling and contaminant residues in top predators while adapting to contaminants. All of these affect ecosystem services.

Invasive species impacts are expected to increase as a result of global climate change. For instance, increased water temperature will alter thermal habitats and the potential range expansion of aquatic species, e.g., northern fish populations may be threatened by range expansions of warm water, southern fish populations. Some aquatic habitats, such as alpine lakes, where resident communities would have to disperse northward over vast differences to colonize colder alpine lakes, may see complete replacement of aquatic communities after extinction events.

Changes in parasitic infestations will also occur as range extensions follow warming trends. For instance, the relatively recent (early 1990s) expansion of the eastern oyster (*Crassos-trea virginica*) parasite, *Perkinsus marinus*, into the northeastern United States appears to be due to warming trends. Invasive species can also be disease vectors.

Successful invasions will change trophic interactions, which may have positive or negative effects on such invasions, with additional effects related to climate change. For instance, the native Atlantic crab, *Carcinus maenas*, preferentially feeds on the native mussel, *Mytilus galloprovincialis*, rather than on the New Zealand invasive mussel *Xenostrobus securis*, because it is easier to handle and break and thus energetically more favorable; this preference only increases with increased temperature.

Determining the ability of invasive species to not only colonize different habitats but also to cause ecosystem impacts as defined by Simberloff, above, requires consideration of 4 factors: arrival, survival, establishment, and spread. Risk assessments of invasive species must consider each of these factors, including environmental variability that can lead to microhabitats, which can support invasive species when the majority of available habitat cannot.

As mentioned previously, habitat change is arguably a greater agent of change than species invasions. Together, these 2 stressors result in complex nonadditive effects that will change ecosystem structure and function. For instance, loss of native habitat will favor invasive species over native species; even a comparatively small increase in habitat changeover time can lead to an abrupt increase in invader abundance. However, the effects of invaders can change over time, modulated by 4 factors: changes in the invasive species; changes in the biological community that is invaded. (e.g., changes in community composition and in individual species characteristics); cumulative changes in the biotic environment that is invaded. (e.g., via feeding or engineering activities); and, interactions between the invading species and other variables that control the ecosystem. (e.g., fire regime in terrestrial ecosystems, hydrology in aquatic ecosystems)

Both positive and negative changes are possible. The following examples are provided by Pejchar and Mooney. An invasive tree in Florida (*Melaleuca quinquenervia*) has a positive effect on honey production but a negative effect on tourism. The introduction of brush-tailed possums (*Trichosurus vulpecula*) to New Zealand resulted in massive defoliation but was highly profitable to the "eco-friendly" fur industry. In South Africa, invasive *Acacia* and *Pinus* species have resulted in reduced stream flow and increased fire intensity, but they have also been positively incorporated into local livelihoods, providing materials for thatching, timber, medicine, charcoal, and firewood. Carbon storage capacity has been lost from the Brazilian Amazon as fire-prone nonnative pasture grasses have replaced rainforest; however, locals have benefited economically in the short term. Thus, apparently harmful effects to biodiversity may not similarly translate into universally negative effects on the well being of humans, particularly over different time scales.

　　Ecosystems will change as a direct (e.g., warming) and indirect (e.g., invasive species) result of global climate change. Such change is inevitable and will occur as organisms adapt to and cope with changing environmental conditions. In the context of global climate change, irreversible changes will become the norm. Risk assessments of invasive species and of other changes enhanced by global climate change should not be based on comparisons to the "status quo," but rather comparisons to the optimum level of ecosystem services possible under changing climatic conditions resulting in changing ecosystems. They should also be based on both negative and positive impacts over different time scales. (Chapman, 2012)

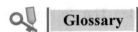

Glossary

invasive species　入侵物种
habitat　生境
relative abundance　相对丰度
ecosystem service　生态系统服务
global climate change　全球气候变化

aquatic community　水生群落
microhabitat　微生境
terrestrial ecosystem　陆地生态系统
biodiversity　生物多样性

Part II Promotion Components
提升篇

This part mainly introduces the latest research contents and directions about ecology in term of the hierarchical structure of biological systems. The selected contents are mainly from the high level journals related to ecology such as Science, Ecology Letters, Global Change Biology, and are sorted and adapted according to the needs. The part content can not only improve readers' reading ability, but also make readers understand the frontiers of ecology studies.

Unit 6   Studies on autecology ecology

Lesson 22   Growth and yield stimulation under elevated CO₂ and drought: a meta-analysis on crops

Atmospheric CO_2 concentrations have increased since the onset of the industrial revolution and are expected to further increase, together with more frequent periods of low water availability (IPCC 2014). Drought stress is a major constraint to crop yield and quality. Therefore, in order to maintain future food security, knowledge about the interactive effects of elevated CO_2 (eCO_2) and drought stress on crop performance is of great importance. In many regions climate change will not only lead to enhanced drought stress via reduced rainfall but also to lower air humidity due to higher temperatures, leading to an increased leaf-to-air vapor pressure deficit. Numerous studies dealt with effects of eCO_2 or drought stress on plant performance. Fewer studies addressed the complex interactive effects of eCO_2 and drought stress. Most studies that examined the effect of low water availability in combination with eCO_2 concluded that eCO_2 leads to decreased plant transpiration, and therefore eCO_2 is expected to ameliorate the adverse effects of drought stress.

Plants with different mechanisms of carbon fixation are expected to respond differently to eCO_2 and drought. In terms of stimulation of photosynthesis and growth, C4 plants generally respond less to eCO_2 than C3 plants. This difference is due to the efficient way of CO_2-fixation of C4 plants, which have a cellular mechanism to increase CO_2-concentration in the bundle sheath cells where the carbon fixing enzyme rubisco is located. Because of this cellular carbon concentrating mechanism, C4 plants are able to reduce their stomatal conductance to a minimum, especially when water loss is high. This leads to a higher tolerance to drought of C4 over C3 species and to different CO_2-response curves of photosynthesis.

Drought and eCO_2 have separate as well as combined effects on plant biomass and yield. Biomass is expected to increase under eCO_2 due to greater carbon fixation and due to a prolonged active growing period by maintaining the soil water reserves for longer. An extended period of grain filling under eCO_2 may subsequently increase crop yield, because the efficiency of carbon allocation to grains and the filling period duration affect yield quantity. Though, whereas biomass production is the direct result of photosynthetic carbon assimilation and water uptake, the formation of yield is a more complex process with additional physiological mechanisms involved. Hence, the effects of eCO_2 and drought on biomass and yield might not be identical.

A meta-analysis is a powerful tool to quantitatively evaluate the overall pattern across many independent studies because it combines the results into one measure: the effect size. Meta-analytical studies on the interactive effects of eCO_2 and stomatal conductance or seasonal drought have been conducted; however, the interactive effects of eCO_2 and experimentally imposed drought (i.e., studies that simultaneously applied at least two water treatments) have only been touched on briefly in meta-analyses, or considered only C3 crops.

Van der Kooi et al. focused on plant species that are important for agricultural uses, i.e., crops and pasture grasses, for different reasons. Identifying the effects of global change on crop growth and yield is essential in order to maintain future food security and crops can serve as an exemplary model for other plant species. From a quantitative perspective, crops are particularly suitable because many experimental studies examined the interactive effects of eCO_2 and drought in this group of species, allowing comparisons between species and functional groups. For their meta-analysis, they analyzed studies that juxtaposed an experimental drought treatment with well-watered controls in interaction with exposure to elevated and ambient CO_2. They evaluated the differences between functional groups such as C3 and C4 or annual and perennial crops. Where sufficient data were available, also the differences between important crop species and between different experimental setups were studied. They aim to test the hypothesis that crops and pasture grasses benefit more from eCO_2 under drought than under well-watered conditions. (van der Kooi et al., 2016)

Lesson 23 Mycorrhizal associations of trees have different indirect effects on organic matter decomposition

Soils represent the largest reservoir of terrestrial carbon, and decomposition is the main process that depletes this C pool. Mycorrhizal symbioses (i.e. Arbuscular mycorrhizal(AM) vs. ectomycorrhizal(ECM)) of plants indirectly affect decomposition, but they do so in distinct ways, and this may result in divergent patterns of C loss. AM and ECM plants promote soil microbial communities with distinct functional attributes, and they tend to generate litter of differential quality. However, it is unknown whether these factors individually, or together, result in mycorrhizal-specific patterns in organic matter decomposition. Such knowledge is critical for resolving why AM-dominated biomes and ecosystems tend to contain lower stocks of soil C (or lower soil C: N) relative to their ECM counterparts.

Mycorrhizal symbioses of plants differ in the activity and composition of the decomposers they promote, and this may relate to their distinct strategies for nutrient acquisition. AM fungi "scavenge" mineral forms of N and phosphorus and may stimulate decomposers to increase both C and N mineralization. In contrast, ECM fungi can "mine" N directly from organic matter via secretion of extracellular enzymes, which creates a competitive context between mycorrhizal fungi and decomposers for organic N. These mycorrhizal strategies may lead to different assemblages of saprotrophs in AM and ECM-dominated ecosystems and could explain why leaf-litter decay rates are often greater in AM-dominated versus ECM-dominated forests.

The quality of plant litter may also promote distinct mycorrhizal effects on decomposition. Arbuscular mycorrhizal plants tend to produce leaf litter with a higher quality (i.e. lower C:N) and greater decomposability relative to ECM plants in temperate ecosystems, although it remains unclear if this pattern holds across all ecosystems and plant lineages. Root litter is more poorly studied, but the available evidence suggests that ECM roots and fungi are recalcitrant and may decompose more slowly than their AM counterparts.

The mycorrhizal identity of a plant may affect how its litter interacts with existing decomposers and soil organic matter pools. Litters can decompose more rapidly in "home" soils relative to "away" soils, but this can depend on the ecosystem type and its mycorrhizal composition. For example, decomposition can be faster in AM-versus ECM-dominated forests, and AM litter decay can be amplified in AM soils. However, it is unclear how these patterns reflect how new litter C interacts with decomposers and organic matter, and specifically, whether new C adds to, or accentuates loss of, existing organic matter in the soil profile. This issue

Unit 6　Studies on autecology ecology

may be most relevant for fine roots, which turn over directly in mineral soils and are a major source of soil C. If the relationships among litter, decomposers and organic matter differ by plant mycorrhizal identity, they may contribute to patterns in soil C balance across terrestrial ecosystems.

In the experiment performed by Taylor et al., they sought to understand how the mycorrhizal identity of trees affected interactions among fresh litter, decomposers and soil organic matter, with a specific focus on C loss via heterotrophic respiration. To isolate effects of mycorrhizal fungi from those of climate and soil type, they conducted an experiment in a temperate forest ecosystem where AM and ECM trees naturally co-dominate. They first characterized soil biogeochemical patterns (total soil C and N, soil microbial biomass C, N and P, and the availability of N and P of individual species of AM and ECM trees *in situ*. They also characterized differences in soil pH between AM and ECM trees, as pH can be a determinant of soil chemistry and tends to be lower in ecosystems dominated by ECM versus AM plants. They then tested the importance of the mycorrhizal identity of mineral soil (including its natural organic matter and decomposers) and litter on soil C loss through respiration. To this end, they collected mineral soils and litters from species of AM and ECM trees in the field and conducted a microcosm experiment in the laboratory. They quantified heterotrophic respiration of soils and determined the response of respiration to additions of leaf and root litter. By isolating soils from plants and mycorrhizal fungi, they specifically determined how indirect effects of mycorrhizal identity contribute to organic matter decomposition.

Taylor et al. hypothesized that AM trees would demonstrate lower soil total C, N and C:N, lower microbial biomass C:N, but greater soil pH, and available N and P relative to ECM trees, as a reflection of their nutrient acquisition strategies and subsequent effects on soil chemistry. For the microcosm experiment, they hypothesized that heterotrophic respiration would be greater in AM soils vs. ECM soils, and that the addition of AM litter would stimulate respiration more than ECM litter. These expectations were based on the idea that AM-dominated ecosystems promote litter decomposition relative to their ECM counterparts and that AM plants tend to produce more labile litter than ECM plants. (Taylor et al., 2016)

Lesson 24   Effects of enhanced UV-B radiation on the nutritional and active ingredient contents during the floral development of medicinal chrysanthemum

The reduction of stratospheric ozone has become a global problem, which results in the increase of solar UV-B radiation on the earth's surface. Enhanced UV-B radiation has already got comprehensive attention among the governments and scientists. In general, enhanced UV-B radiation can negatively affect growth, physiology and productivity of plants. In recent years, however, some interesting results about UV-B beneficial effects on secondary metabolism processes in plants have been found. Most of the medically active ingredients in medicinal plants are secondary metabolites. Therefore, it is considered of interest to determine if active ingredient contents in medicinal plants can be improved by enhanced UV-B radiation which is a simple and environmental-friendly method.

At present, a few studies have reported UV-B effects on medicinal plants. Previous results indicated that enhanced UV-B radiation could induce secondary metabolism processes, and increase active ingredient contents in medicinal plants. However, enhanced UV-B radiation in previous experiments was applied during growth stages of plants, which was difficult to manage (especially for regions with no reliable electricity source) and required a large investment in production. Therefore, further studies to explore methods for improving active ingredient contents by UV-B radiation were deemed necessary.

In order to better applying UV-B radiation technology on medicinal plants, some scientists began to study the effects of UV-B radiation on isolated organs of medical plants. Schreiner et al. reported the effects of short-term and moderate UV-B radiation on active ingredients in different postharvest organs of nasturtium (*Tropaeolum maius* L.). The results showed that enhanced UV-B radiation increased the glucotropaeolin concentration up to 6-fold in comparison to the control plants. Sun et al. found that UV-B radiation increased the contents of secondary metabolites in postharvest leaves of ginkgo (*Ginkgo biloba* L.). In previous experiments, Yao et al. also found that moderate UV-B radiation could affect biochemical traits in isolated flowers of medicinal chrysanthemum. To our knowledge, there have been limited efforts to know effects of enhanced UV-B radiation on the contents of nutritional and active ingredients during the floral development of medicinal chrysanthemum. In addition, the scientists are just beginning to study UV-B effects on isolated organs of medicinal plants. So, more works should be done for better evaluation of the application of UV-B radiation in medicinal plants.

Medicinal chrysanthemum (*Chrysanthemum morifolium* Ramat) is one of important export medicines in China. Medicinal chrysanthemum flowers are used in traditional medicine where they play an important role in improving liver function, decreasing inflammation, improving eyesight and serving other anti-inflammatory detoxification roles. Flavonoids and chlorogenic acid are main active ingredients in chrysanthemum flowers, and free amino acid, vitamin C and soluble sugar are main nutritional ingredients in flowers. The paper mainly studied the effects of enhanced UV-B radiation on biochemical traits and quality during the floral development of medicinal chrysanthemum, in order to finding the stage which is more sensitive to enhanced UV-B, so as to determining the best harvest stage according to the contents of nutritional and active ingredients in flowers. This will be helpful for further research about the application of UV-B radiation on medicinal plants. (Ma et al., 2016)

Lesson 25　Interactive effects of temperature and pCO_2 on sponges: from the cradle to the grave

The Intergovernmental Panel on Climate Change predicts that global warming resulting from increasing atmospheric carbon dioxide levels ($CO_{2\ atm}$) will see global mean sea surface temperatures increase 1.1–4.0 ℃ by the end of this century. Further to this, as the partial pressure of CO_2 (pCO_2) in sea water increases with increasing $CO_{2\ atm}$, the pH of the world's oceans is predicted to decrease 0.2–0.32 units by 2100 (IPCC, 2014), in a phenomenon referred to as ocean acidification (OA). Increasing evidence from meta-analyses demonstrates that multiple climate stressors have greater deleterious impact on marine organisms than stressors applied in isolation. However, compared to the number of single stressor studies, few experimental studies have considered the combined effects of ocean warming (OW) and OA, or the impacts on different life-history stages of the same species. Furthermore, the major research focus in tropical marine ecosystems has been on reef-building corals, where a bleak future is being forecast for a group of organisms already experiencing global declines in abundance.

Where coral abundance has already declined, "winners" and "losers" are beginning to emerge. While most regime-shifts away from coral-dominated states are to macro-algal dominated reefs, other states are also possible with sponges being identified as potential winners in the face of environmental change. Increasing sponge abundance has been reported from coral reefs globally, likely as a result of decreased spatial competition with corals. Recently proposed positive feedbacks resulting from dissolved organic carbon cycling between seaweeds and sponges suggest that these alternate states are also likely to persist, as sponge and seaweed growth are enhanced to the detriment of coral recovery.

Sponges form an important functional component of coral reef ecosystems, with efficient water filtration rates and high particle retention efficiencies that make them a critical link in the coral reef food chain. This link has recently been reinforced by the discovery of the "sponge loop", which demonstrates how sponge-facilitated carbon flow contributes to the productivity of oligotrophic coral reef waters. In addition, coral reef sponges contribute to reef primary productivity and nutrient cycling through their associated microbial symbionts, are involved in reef bioerosion and consolidation and are strong spatial competitors. Given these important functional roles, any change in sponge population dynamics and abundance could have significant flow-on consequences for coral reef function and health.

Sponges should be particularly vulnerable to alterations in water chemistry as just one cell layer separates a sponge from the external environment. Therefore, changes to ambient water temperature and pH may strongly influence cellular processes, especially considering the control that both temperature and CO_2 exert over fundamental metabolic processes. Small increases in temperature can lead to elevated metabolic rates, while acid-base imbalances under high CO_2 levels can depress metabolic activity, both of which lead to energy being diverted away from important metabolic and ecological processes. Basal marine invertebrates such as sponges are thought to be particularly susceptible to the effects of

OA due to their limited capacity for acid-base regulation; however, little is known about how sponges respond to future OW and OA scenarios.

Studies focussing exclusively on the effects of OW have demonstrated that some sponges are sensitive to temperatures predicted under future climate change scenarios. For example, the Caribbean sponge *Xestospongia muta* exhibits elevated expression of heat shock proteins and mortality when exposed to 30 ℃ and the Great Barrier Reef sponge *Rhopaloeides odorabile* demonstrates a thermal threshold of 32 ℃ with a breakdown in host-symbiont molecular interactions occurring at elevated temperature. Importantly, the sponge holobiont can be categorized into two nutritional types that have been shown to exhibit contrasting responses to temperature anomalies: (i) phototrophic species, which are those where >50% of energy requirements are acquired from photosynthetically fixed carbon; and (ii) heterotrophic species, which are those primarily reliant on suspension feeding for carbon requirements.

Studies focussing exclusively on the effects of OA at volcanic CO_2 seep sites have reported contrasting results. While sponge diversity was found to increase at low pH sites (pH 7.8–7.9) in the Mediterranean, sponge diversity decreased at low pH sites (pH 7.73–8.00) in Papua New Guinea, although particular species did become significantly more abundant at sites with active CO_2 bubbling. Although these studies provide some insight into sponge responses to a changing climate, the co-occurrence of OW and OA requires studies that consider these factors concurrently.

Research into the combined effects of elevated temperature and pCO_2 on sponges has predominantly focused on the erosion rates of bioeroding sponge species, with bioerosion generally increasing under future OW and OA scenarios. In addition, research exposing six sponge species to combined OW and OA (31.5 ℃/pH 7.8) found no impact on growth, survival and secondary metabolite biosynthesis. Furthermore, despite slight negative effects of elevated pCO_2 on spicule biomineralization, OW and OA ($\sim$27℃/$pCO_2\sim$1,100 ppm) had little effect on overall survival or growth rates in *Mycale grandis*. Finally, in *X. muta*, phototrophic cyanobacterial symbiont productivity was reported to decline with exposure to elevated temperature and pCO_2 (31.4℃/$pCO_2\sim$800 ppm), although no evidence of bleaching or associated host stress was reported despite a reduction in holobiont carbohydrate levels and reduced stability of the sponge microbiome. While these studies suggest a degree of sponge tolerance to climate change scenarios predicted for the end of this century, they provide little insight into the effects of environmental change at the cellular level required for a mechanistic understanding, especially when stressors are impacting in synergy.

The survival of early life stages is fundamental for population persistence, yet no studies have explicitly tested the early life-history response of sponges to the combined effects of OW/OA. In the only study that considered the impact of climate change on different sponge life-history stages, larvae of *R. odorabile* were found to have a much greater thermal tolerance than adults. However, in many marine invertebrates, OW and OA are considered to have deleterious impacts on the success of early life stages.

In the study performed by Bennett Bennett et al., they examined the physiological responses of four abundant Great Barrier Reef (GBR) sponge species—the phototrophic *Carteriospongia foliascens* and *Cymbastela coralliophila* and the heterotrophic *Stylissa flabelliformis* and *Rhopaloeides odorabile*—to a rise in pCO_2 and sea water temperature over 3 months, to address the hypothesis that phototrophic and heterotrophic sponges will exhibit differential responses to climate change. Observed and measured physiological performance of sponges under OW and OA was assessed to gain a better understanding of sponge energetics and the implications for ecological performance, in a rapidly changing climate. (Bennett et al., 2017)

Lesson 26　Salinity influences arsenic resistance in the xerohalophyte *Atriplex atacamensis* Phil

Arsenic (As) is a highly toxic element for all living organisms and a global concern for human health. High concentrations of As in groundwater, soils and sediments in various parts of the world have been identified. This is especially the case in Northern Chile (regions Arica and Parinacota, Tarapacá and Antofagasta) where As is predominantly released from volcanic rocks, sulphide ore deposits and their weathering products at the Andean volcanic chain. Human activities, mainly copper mining, contribute to As release in the environment. As a consequence, the presence of extremely high As concentration has been reported in local places. In the main river of this area, the Loa river, high concentration of As (average: 1,400 mg L^{-1}) have been recorded. This high As concentration constitutes a major risk for local populations using this water as drinking water or for irrigation of vegetable garden.

Arsenic is a metalloid occurring predominantly in inorganic form as oxidized arsenate (As(V)) and reduced arsenite (As(III)). Arsenate is an analogue of phosphate, competing for the same uptake system and can thus disrupt phosphate metabolism and substitute phosphate in ATP. In plants, arsenate may be reduced by non-specific arsenate reductase to arsenite which tends to be complexed with thiol rich-peptides and stored in vacuoles. Arsenite binds to sulfhydryl groups of enzymes and proteins with subsequent inhibition of cellular functions. As(III) is transported across biological membranes in the neutral form through nodulin 26-like intrinsic proteins (NIPs) aquaporin channels. Arsenic toxicity is also mediated by oxidative stress through generation of reactive oxygen species (ROS) and inhibition of antioxidant defences in plant tissues. Environmental constraints induce modifications in the plant hormonal status. Stress-induced ethylene oversynthesis may hasten the leaf senescence ultimately leading to plant death.

Some plant species may help to reduce the risk of As dispersion from contaminated sites and stabilize substrate through avoidance of erosion processes. The fern *Pteris vittata* is a well known hyperaccumulator able to accumulate more than 4,000 µg g^{-1} of As in its aerial part. In Chile, however, areas affected by As contamination are also extremely arid, and *P.vittata* could not be considered as a good candidate for phytostabilization purposes. In contrast, the local *xerohalophyte* plant species *Atriplex atacamensis* has been recommended as a promising plant species for phytomanagement of As contaminated sites. *Atriplex atacamensis Phil*. is native of Northern Chile (Atacama desert) and is able to cope with high As(V) concentration (up to 1,000 µM As in nutrient solution). Tapia et al. recently confirmed that *A. atacamensis* is highly resistant to As contamination in soils and does not exhibit toxicity symptoms when growing on a Pre-Andean soil contaminated with more than 100 mg As kg^{-1}.

In its natural environment, *A. atacamensis* is not only exposed to high external As doses but also frequently encounters extremely high salinities. The impact of salt on heavy metal accumulation has been extensively documented in the literature. Salinity was reported to affect pollutant absorption and translocation on the one hand, and to impact the physiological strategies of the plant to cope with the accumulated heavy metals on the other hand. In contrast, data concerning salt effect on As uptake and accumulation by plants remain scarce. Plant tolerance to salinity may rely on the synthesis of protecting compounds such as proline, glycinebetaine, trigonelline and polyamines involved in osmotic adjustment and/or protection of cellular structures. These compounds, however, are not specific to NaCl constraint and both polyamines and glycinebetaine were reported to be involved in *A. atacamensis* resistance to arsenic. It may thus be hypothesized that physiological strategies triggered by NaCl on the one hand and As on the other hand may overlap to some extent. However, there is a crucial lack of information concerning the plant behaviour when both constraints are present concomitantly.

The physiological consequences of As accumulation is a direct function of As speciation within plant tissues but also As distribution between symplasm and apoplasm. No data are available on the putative effects of salinity on these parameters. Moreover, the plant response to abiotic stress is not only a function of stress intensity but also a direct function of the total duration of stress exposure. In the presented work, the tested hypothesis are (i) that NaCl may influence *A. atacamensis* response to As toxicity, and (ii) that such an influence may vary with the duration of the treatment. For this purpose, physiological parameters were quantified in plants exposed for 2 and 4 weeks to NaCl, arsenate or arsenate + NaCl in relation to As accumulation, speciation and distribution in this xerohalophyte plant species. (Vromman et al., 2016)

Unit 7　Studies on population ecology

Lesson 27　Climate, invasive species and land use drive population dynamics of a cold-water specialist

  Many cold-water fishes are predicted to experience range contractions as a result of climate change. These projections are particularly concerning given that freshwater biota are subjected to elevated rates of extinction compared to other taxa. However, most studies describing climate effects on freshwater fishes have largely focused on quantifying how species distributions are related to temperature, ignoring other sources of environmental variation that play important roles in determining contemporary population dynamics. Accordingly, the cumulative, population-level effects of climate and other anthropogenic stressors are largely unresolved for freshwater fish. Focusing on the direct effects of climate alone may lead to conservative predictions that underestimate climate impacts if stressors are synergistic, or misappropriation of resources towards climate mitigation before more immediate threats are appropriately addressed.

  Although patterns in species occupancy or physiological tolerances can help inform existing research gaps, demographic data collected across time and space are particularly powerful for directly quantifying the effects of climatic variation on populations relative to other intrinsic and extrinsic sources of variation. Identifying drivers of spatio-temporal population dynamics also provides an opportunity to verify predictions based on patterns in species occupancy, and a means to prioritize management and conservation actions on the contemporary time-scales and units of measurement (e.g. abundance) that underlie many fisheries and wildlife management programmes.

The uncertainty surrounding how climatic and non-climatic stressors additively or interactively (i.e.synergistically) influence population dynamics in freshwater fishes partly reflects a lack of spatially replicated long-term data. Nevertheless, monitoring programmes are frequently used to assess the status of many fishes, and these data can be used to describe key drivers of population persistence, including average abundance (N), population stochasticity (σ) and population growth rate (r). Spatial patterns in N, r and σ can be used to quantify the cumulative effects of multiple stressors and help identify major population drivers.

Bull trout (*Salvelinus confluentus*) —a threatened species protected under the US Endangered Species Act in the United States (US Office of the Federal Register 1998) —are among the most temperature-sensitive freshwater species in North America, making them vulnerable to climatic warming. Furthermore, their imperilled status is attributed to negative interactions with invasive fish species and habitat degradation. To date, no studies have quantified how climate and other human stressors influence bull trout population dynamics in diverse abiotic and biotic environments.

As a case study describing how threat assemblages influence population dynamics in freshwater ecosystems, Kovach et al. used temporally and spatially extensive bull trout monitoring data to address three questions: (i) Do climate, invasive species and land use act additively, interactively or antagonistically to influence N, σ and r? (ii) Which stressors have the largest effect on population dynamics? And (iii) how do life-history variation, physical habitat and management action influence bull trout relative to these stressors?（Kovach et al., 2017）

Lesson 28 Does movement behaviour predict population densities? A test with 25 butterfly species

Diffusion is a leading model of animal movement in ecology. Theoretical ecologists use diffusion processes to model animal movement. Empirical ecologists and conservation biologists use diffusion models to scale up from small-scale measurements to large-scale generalizations about long-distance movement, dispersal and (re)colonization processes. Diffusion approximates a correlated random walk, so it can be explicitly estimated from animal tracking studies, or by embedding correlated random walk (CRW) models into spatially explicit individual based models.

The use of diffusion to model animal movement has been widely criticized as too simplistic. This is because animals vary in their movement behaviour based on sex, habitat association, habitat quality, proximity to boundaries or resources, age, propensity to breed, proximity to predators and numerous other factors. In addition, diffusion models assume that each move step is independent of the prior step, that is, they are Markovian, but memory and perceptual abilities in animals make it likely that more complex processes regulate animal movement.

If diffusion is an adequate model of animal movement then slower diffusion rates should lead to higher abundances. However, there are reasons why we might not see this relationship in real landscapes. First, it may be that diffusion is an inadequate model of movement, which would make diffusion rates poor metrics for predicting abundance. Second, even if diffusion is generally an appropriate model, abundance could primarily be determined by responses to patch edges rather than movement within a patch. Third, if the time-scales of population or habitat dynamics are similar to those of movement, vital rates and/or metapopulation dynamics, rather than movement, may be the primary predictors of abundance.

To date, several studies have used diffusion rates to predict animal densities. Among case studies, seven out of nine species show the qualitative pattern of higher densities in areas with slower estimated diffusion rates, and two do not. In addition, some studies only use simpler metrics such as move length, tortuosity, total distance moved and speed to describe movement, for example, based on mark-recapture, or radio or satellite telemetry data. These studies often cite the theoretical literature from models based on CRWs or diffusion, but make inferences about animal movement based on 1–2 of the aforementioned simple metrics (e.g. move length or distance moved) rather than formal diffusion-based metrics.

In the experiment performed by Schultz et al., they test whether slower movement generally corresponds to higher abundance across 25 butterfly species in four habitat types. They use fine scale behavioural data to make predictions across broad landscape and community scales, extending past the generality of previous studies that generally include only one or two species in no more than two habitat types. In addition, they test whether simpler summary statistics (move length, move time, turning angle and expected net squared displacement) are sufficient as surrogates of diffusion to predict variation in butterfly densities. They study differs from past tests of diffusion models in that they ask whether these simple models make statistically significant predictions about abundance across a larger number of land cover classes than past studies, and whether the predictive ability is consistent across multiple species within a taxonomic group. （Schultz et al., 2017）

Lesson 29 Anthropogenic-driven rapid shifts in tree distribution lead to increased dominance of broadleaf species

The distribution of forests is primarily determined by climate, and temperature is one of the most important factors limiting the geographical range of species at large scales. Shifts in poleward and altitudinal ranges have been interpreted as the fingerprint of recent climatic warming on the biosphere. Numerous studies have already reported shifts in the geographic distribution of plant species, presumably due to climate change, especially in response to recent warming. These shifts have been observed frequently in mountains, implying movements uphill, whereas documented poleward shifts remain relatively scarce. Shifts in the range of species associated with 20th century warming have been well documented for a wide range of taxa. Many species are not migrating fast enough to keep pace with the rapidly changing climate and in consequence are becoming vulnerable. This is especially true for trees, which are sessile, long-lived and often slow to mature. As a result, there is a growing concern that anthropogenic global change (including biological invasions, land use changes and climate change) may alter the distribution, composition and function of forests and associated functioning.

Shifts in the range of species distribution require that mortality and recruitment are not in balance over time. An imbalance might occur due to increased tree mortality leading to local extinction at the boundaries of the species geographic distribution, due to colonization at the leading edge of the species' range, or both. Trees depend on seed dispersal and seedling colonization of newly favourable areas to respond to climate warming and on mortality and extinction rates in less favourable areas; both or one of them needs to occur over several generations. The magnitude of range shifts in response to particular warming scenarios is often predicted using the correlation between current ranges of species and current climate, under the assumption that the distribution of species will shift in synchrony with changing climate.

However, there are several reasons why changes over time due to changing environmental conditions may not mimic spatial responses to the same environmental factor (i.e. space for time substitution may not hold). A key aspect is the large impact of nonclimatic factors on species' performance. For instance, many types of land use change, and limits to dispersal as a consequence of physical barriers, fine-scale habitat heterogeneity and changes in grazing pressures are also important in determining the actual or potential distribution range of species. All these factors often act in opposite ways, making it difficult to understand the significance and direction of migratory pressures. Over the last three decades, the cessation of forest management is among the most important nonclimatic changes in Spain and in most of European forests. This cessation has led to a gradual increase in forest density and changes in composition, structure and demographic processes.

In the experiment performed by Vayreda et al., they assess for the first time the combined effect of changes in climate (e.g. recent climate warming), forest management and disturbances (e.g. wildfires) on the geographic distribution of the most abundant forest tree species in Peninsular Spain. This Mediterranean region has a long history in the management of forests. They use about 33 000 forest plots from the Spanish national forest inventory (IFN) which were resampled 10 to 12 years apart over the period 1986–2008. In addition, they explore the links between these changes and the functional traits at the level of species and families (i.e. *Fagaceae* and *Pinaceae*, hereafter, broadleaves and conifers, respectively). They further investigate the habitat preferences of species to shed light on the ecological mechanisms driving the shifts and the implications of those shifts under future environmental change. (Vayreda et al., 2016)

Lesson 30 Species' traits influenced their response to recent climate change

The rate of warming over the past 50 years (0.13℃±0.03℃ per decade) is nearly twice that for the previous 50 years, and the global temperature by 2100 is likely to be 5–12 standard deviations above the Holocene mean. The effects of climate change on some species are already being witnessed, with changes documented in spatial distribution, abundance, demography, phenology and morphology. However, to date, no quantification of the number of species for which at least one population has been currently impacted by climate change, and the extent of these impacts, has been conducted, even for the better-studied taxa such as birds and mammals. The predominant focus of climate change assessments for species has been that of bioclimatic niche modelling, which focuses on correlative analyses between species' geographic ranges and bioclimatic variables, but these studies ignore observed changes in distribution, phenology and abundance of species in response to contemporary climate change. Species' life-history traits, such as dispersal and generation length, have been hypothesized to be important in determining species' sensitivity to climate change and their capacity to adapt to it, but only a limited number of studies have so far provided evidence that animal species with certain traits are more likely than others to be adversely affected by changes in climate.

In the study performed by Pacifici et al., they first aimed at performing a meta-analysis to identify the life-history traits that confer vulnerability to climate change in birds and mammals. From a literature search, they identified 70 studies covering 120 mammal species and 66 studies relating to 569 bird species whose populations had (or sought evidence for) a response to climate change in recent decades. They divided this response into four categories: negative, if >50% of the populations experienced reductions in one or more of the following parameters: population size, geographic range size, reproductive rate, survival rate, body mass; positive, if the species experienced increases in one or more of the parameters and/or adaptability to new climatic conditions; unchanged, if no response was observed despite the recorded change in climate; and mixed, if the species showed opposite responses of one or more of the parameters across its geographic range. For all mammals and birds covered by the studies, they compiled data on selected intrinsic traits and spatial traits to assess quantitatively which of these are associated with negative responses to climate change. To control for the magnitude of climate change experienced, they also computed the mean difference in temperature between the present and the recent past within the geographic range of each species, treating breeding and non-breeding ranges separately for migratory birds.

By using information on the impacts of climate change in the study areas and life-history traits, Pacifici et al. were able to identify the species whose populations are more likely to have experienced negative impacts in the regions affected by climatic changes as those described in the analysed papers. They estimated the likelihood of a species' population to have exhibited any of the four categories of responses to climate change with a multinomial regression model. This allowed us to test their hypotheses about the relationship between intrinsic and spatial traits and the responses of mammals and birds to climate change. Since they believe that these factors mediate the response to climate change similarly worldwide, although future studies will be crucial to test this assumption, they then predicted the likely past responses of all birds and terrestrial non-volant mammals listed as threatened in the International Union for Conservation of Nature (IUCN) Red List of Threatened species. By making predictions on the species for which the levels of climatic hazard experienced are known, they provide the first quantification of the number of taxa that may have already been impacted, although further data need to be collected to say with certainty that there has been an effect on the whole species' persistence. They focused on threatened species because the vast majority are known or inferred to have declined; therefore, if they are at risk from climate change there is a real chance that it has played a role in these declines, even if it was not recorded in the assessments.

For the first time, Pacifici et al. identified a relationship between a set of several variables, both intrinsic and spatial, and the response of mammals and birds to climate change, whereas previous studies mostly focused on a few biological traits and their relation with the type of impact. In addition, they were able to provide insights into the estimation of climate change threat for poorly studied species. (Pacifici et al., 2017)

Lesson 31　Genetic diversity affects the strength of population regulation in a marine fish

The dynamics of a population are largely a reflection of how the population grows when small and shrinks when large. This type of negative feedback, referred to as density-dependent population regulation, is a prominent feature for the vast majority of wild populations. Although regulation is common, our understanding of this process is far from clear. In the wild, density-dependent regulation is often a nonlinear process that may lead to complex dynamics. Additionally, the degree of regulation can vary widely, resulting in substantial differences in the dynamics of populations. Even within a single species, local populations can exhibit considerable variation in their dynamics, with some populations fluctuating extensively while others remain stable.

Population regulation may result from competition and/or predation. In many cases, intraspecific competition is the underlying cause of density dependence in demographic rates. For example, at high densities, a relative shortage of resources can reduce average rates of reproduction, growth, and/or survival. In other cases, predation is the primary cause of density-dependent regulation. This may occur if predators consume disproportionately more prey when preys are abundant. Predation can also act in concert with competition, if competition makes prey more susceptible, and predation is the proximate source of mortality. Population regulation is itself a dynamic process and the strength of density-dependent regulation may vary if the underlying mechanisms (competition and/or predation) are modified by other factors. For example, the abundance of interspecific competitors can have an effect on the carrying capacity of an environment and thus affect how strongly populations are regulated. Similarly, variation in food availability can affect the intensity of competition. Density-dependent predation can be modified by availability of refuge space for prey and/or the local abundance of predators, especially when interactions among predators inhibit or facilitate the consumption of prey.

Exogenous factors that modify density-dependent interactions are important and worth accounting for, but endogenous factors may also have important effects on dynamics. In particular, the role of genetic/phenotypic variability within a population needs to be considered more carefully. It is well documented that individuals within a population can use different sets of resources consistently and thereby occupy different niches. Moreover, there are many examples where individual niche specialization has a confirmed, genetic basis. If there is link between genotype/phenotype and how individuals use their environment and interact with one another, then this may have important consequences for population dynamics. Mathematical models that describe competition between two individuals as a function of their phenotypic similarity suggest that populations with greater genetic and phenotypic diversity may, on the whole, interact less strongly, leading to weaker population regulation and greater abundance in the long term. Recent applications of similar models indicate that phenotypic variation can have a major effect on demographic rates. For example, within-cohort variation in phenotypic traits such as size and growth can be responsible for more than half the observed variation in mortality rates of juvenile fishes.

Although theory suggests that the phenotypic/genetic composition of a population can have large effects, understanding the extent to which variation among individuals truly affects the dynamics of populations requires detailed, empirical study. Particularly needed are large-scale studies that assess the effects of genetic variability by comparing the dynamics of multiple populations within a species' range. In the experiment performed by Johnson et al., they examined variation in the dynamics of populations of a live-bearing, marine fish. In collaboration with an organization of citizen scientists, they were able to examine the dynamics of populations throughout much of this species' range (i.e., over 4° of latitude and approximately 700 km of coastline). At these spatial scales, the dynamics of populations varied substantially. The purpose of this study was to evaluate how much of that variation could be explained by differences in genetic diversity within local populations. First, they examined the overall evidence of regulation within populations of black surfperch. Next, they tested the hypothesis that genetic variation within a local population would be related to the strength of population regulation in black surfperch. Finally, they examined whether diversity within a population was related to habitat use and spatial clustering: two related mechanisms that may affect the intensity of competition and, ultimately, the strength of population regulation. (Johnson et al., 2016)

Unit 8　Studies on community ecology

Lesson 32　Levels and limits in artificial selection of communities

Because of the time required for natural selection to occur in nature, artificial selection has been a major argument in the development of the theory of heredity with modification by Darwin, with the deep analysis of pigeon breeding genealogy. In the line of Darwin, several experimental studies have investigated the effects of artificial selection at the group level, to feed the debate on the level of natural selection. Experiments testing group artificial selection involved beetle populations, plant populations, chicken populations, but also two-species beetle communities or multiple-species microbial ecosystems. Recent research also focuses on the ecological consequences of selection of plant trait-associated microbiomes. Consequences of these results for natural selection in nature have already been discussed.

In parallel, much of the work done in modern molecular genetics is focused on the genetic basis of organism phenotypes and changes in alleles frequencies associated with selected phenotypes. However, this cannot be the unique focus when dealing with community or ecosystem artificial selection: in addition to variations in gene frequencies, changes in ecosystem or community phenotype could be due to changes in intraspecies interactions among individuals and species composition. A simulation model even demonstrate that ecosystem artificial selection can occur "without genetic changes", i.e. only because of changes in species composition. Before asking the question of genetic mechanisms involved in the modification of the ecosystem phenotype, it is important identifying the level at which phenotype variance occurs: community, population or individual genes? A first objective of this study was to bring an experimental proof of principle that community structure, especially the structure of interaction networks of communities, are significantly affected during the artificial selection procedure.

A second objective was to document how far human can go in changing ecosystem phenotype by artificial selection. Whereas limits in artificial selection have been well described and formalised at the individual level, the degree to which ecosystem properties may be improved by artificial selection remains unclear. Two different interpretations of the nature of variation in ecosystem artificial selection are leading to opposite predictions on the link between variance and heritability and their consequences on the limits in ecosystem artificial selection.

First, it has been well established that, during artificial selection of individual organisms, directional selection by truncation leads to a reduction in phenotypic variance, depending on the intensity of selection i. $i=z/p$, where z is the ordinate of the normal curve at the truncation point and p is the percentage of selected individuals (or the selection rate). In directional selection, the variance of the individuals selected in the parental generation decreases by a factor of $1-i(i-x)$, where x is the abscissa of the truncation point of the normal curve. Genetic variance can be "used up" by selection in a manner that is proportional to the relative reduction in parental phenotypic variance via the fixation of favourable alleles and the elimination of unfavourable alleles as well as rare alleles by drift (i.e. sampling effect). Exceptions occur when a selected trait involves a very large number of loci. To summarise, a decreased genetic variance should lead to decreased phenotypic variance and consequently to a decrease in heritability and selection efficiency. This could be true for other level of organisation such as the ecosystem, in which genetic variance could be "used up" through the successive loss of rare alleles, individuals or species. In this case, an observed decrease in the variance of ecosystem phenotype should be interpreted as a loss of genetic diversity by sampling effect; ecosystem phenotype variance and heritability should thus decrease along generations and be positively correlated. The limits in ecosystem artificial selection would thus be determined by the initial genetic diversity, size of the population and intensity of the sampling effect.

Another argument leads to an opposite prediction on nature of the limits in ecosystem selection and the sign of the correlation between variance and heritability. According to Lewontin, there are three conditions needed for selection to occur: (1) there must be phenotypic variance among the different individuals experiencing selection; (2) this phenotypic variance must be heritable; and (3) phenotypic differences must be linked with different fitness values. In artificial selection experiments, this third condition is always true, as the breeder/experimenter selects individuals based on phenotypic differences. But Penn et al. pointed out that Lewontin's first and second conditions (variance and heritability) could be at odds as far as ecosystem artificial selection is concerned. Indeed, Swenson et al. did not observe any effect of the size of the sample used to create the offspring generation on ecosystem phenotype variance; they interpreted this result as a proof that the intensity of the sampling effect was not determinant in the variance of the ecosystem phenotype, because its importance should have been lower with large samples than with small ones. Consequently, they proposed that ecosystem variance was determined by the stochastic dynamics of ecosystem or butterfly effect, which occurs whatever the importance of initial differences due to the sampling effect. Penn et al. reported that the stochastic ecosystem dynamics could potentially reduce the heritability of ecosystem phenotypes because it leads to differences between parental and offspring communities. As a consequence, a high variance in ecosystem phenotype due to the stochastic dynamics of ecosystem would be associated with a low heritability and vice versa. If true, a negative correlation should be observed between ecosystem phenotypic variance and heritability. In that case, artificial selection might involve more than a search for ecosystems with desired phenotypic traits; it might also be a selection of ecosystems quickly arriving at stable local equilibria, such that their properties can be reliably transmitted to the next generation. In ecosystem artificial selection, the limits to transmission of selected variations would thus rely on the ability for ecosystem dynamics to reach quickly a stable equilibrium, not on sampling effect and consequences on initial genetic diversity.

In the experiment performed by Blouin et al., they investigated the limits in ecosystem artificial selection to evaluate its potential in terms of managing ecosystem function. By artificially selecting microbial communities for low CO_2 emissions over 21 generations ($n=7,560$), they found a very high heritability of community phenotype (52%). Artificial selection was responsible for simpler interaction networks with lower interaction richness. Phenotype variance and heritability both decreased across generations, suggesting that selection was more likely limited by sampling effects than by stochastic ecosystem dynamics.（Blouin et al., 2015）

Lesson 33 Effects of temperature variability on community structure in a natural microbial food web

Increased mean temperatures due to climate change are having an effect on community-level dynamics in a majority of ecosystems worldwide. Research has already demonstrated that species distributions are being shifted, thus impacting community interactions, mutualistic dynamics, and the functioning of whole ecosystems. In tropical ecosystems, this increase in temperature is thought to be especially problematic because tropical species already live near their upper thermal limits. In temperate and boreal regions, a 1–4°C increase in temperature is, in regard to performance, possibly more beneficial than detrimental for the existing species because ecosystem processes can be accelerated. However, recent climate models have also predicted that climate change will not only involve an increase in mean temperature, but also an increase in temperature variation and a higher probability of extreme events (e.g. drought and extreme maximum and minimum temperatures; IPCC 2013). With this increased temperature variability, species from both temperate and tropical environments are likely to be exposed to conditions beyond their maximum temperature limit. Extreme temperature events may therefore be the most important driving forces for determining climate change-induced community dynamics in many ecological systems. Increased temperature variability could thus have a larger impact on species than an increase in the mean temperature alone.

However, to date, the majority of climate change experiments have increased temperature according to IPCC predictions (increase of 2–4°C, IPCC 2013), while either reducing or keeping the natural temperature variance at the status quo. As a result, the effect of increasing temperature variability on community dynamics remains unclear. It is therefore essential to start to include temperature variation as a treatment in climate change research in order to more fully address the effect that climate change will have on ecosystems. In addition, many studies have been conducted at the population and species level, while information about the impact of climate change on higher organization-levels, such as whole communities or food webs, is lagging behind.

Food webs in freshwater enclosed habitats (i.e. shallow lakes and ponds), in particular, may be especially vulnerable to climate change because these habitats are usually more isolated than terrestrial ecosystems, resulting in species having fewer options for dispersal when water temperature exceeds their temperature limit. Furthermore, with increasing heat and drought levels due to climate change-induced extreme events, the habitable areas (waterbodies) can reduce in size and possibly disappear, resulting in the local extinction of species. Despite their apparent vulnerability to climate change, only one of 21 studies mentioned in the review by Thompson et al. experimentally addressed the implications of extreme events on the communities in these enclosed freshwater systems.

From the experiments that have been published, predictions can be made with regard to the effect of increased mean temperature on community dynamics in enclosed freshwater systems. There will likely be a reduction in trophic levels, a shift in body size towards smaller species, an increase in bacterial densities, and the abundance and composition of species in different levels in a food web will be disproportionately affected. However, it is more difficult to assess the effect of increased temperature variation on communities because its effect is expected to be nonlinear. In terms of diversity, this nonlinear effect can be compared to the intermediate disturbance hypothesis, in which an intermediate level of temperature variation could provide optimal diversity levels, while a low or high level of this variation could reduce diversity. However, the effect of temperature variation on food web dynamics will also depend on the thermal range of the species, which may be difficult to predict based only on body size or trophic-level position.

Furthermore, the successional stage of these communities may also be important for buffering the effect of increased temperature variation. It can be hypothesized that communities in an early successional stage are more likely to be resistant against environmental changes than communities in late succession, as they contain a high proportion of pioneer species that can prosper under a wide spectrum of environmental conditions. These early successional communities also usually harbour more small species compared to later successional stages. These small species can adapt quickly to environmental changes due to higher metabolic rates and fast generation times, making them less likely to be affected by environmental change. Late successional communities, on the other hand, are composed of species that have been selected via biotic and abiotic environmental filters. The species that successfully pass through these filters are the ones that are able to survive in the current temperature regime of the habitat. These species are usually larger in body size and devote more of their energy to competition than early successional species, which may make them less likely to cope with higher variation in environmental conditions.

To better understand how temperature variation will impact enclosed freshwater food webs, and how the successional stage of a community will influence the results, Zander et al. conducted an experiment using the *Sarracenia purpurea* aquatic model system. This food web is typically tri-trophic and is composed of bacteria, protists, and arthropod larvae. In this system, as well as in larger scale enclosed freshwater systems, early successional communities consist of a pool of small-bodied, fast-growing pioneer species which may be physiologically better adapted to rapid temperature changes than the slower-growing, larger-bodied species found in later succession.

To address the importance of temperature variation on food web dynamics, they increased temperature variation while keeping mean temperatures constant in the experiment performed by Zander et al. They accomplished this by manipulating temperature variance as repeated and alternating extreme hot and cold events. These temperatures were within the observed temperature range measured inside *Sarracenia* leaves in the field and have thus been already experienced by the aquatic community. They then measured the impact that increased temperature variation had on the three trophic levels—bacteria, protists, and mosquito larvae—at two different successional stages.

Zander et al. are aware that climate change-related effects of mean and variance interact, but the purpose of this experiment was to determine the magnitude of influence temperature variation could have on food web dynamics. They explored five hypotheses. (i) Extreme temperature variation events will disproportionately increase the density of bottom trophic level organisms, exploring this question by measuring bacterial density. (ii) Increased temperature variation will reduce species diversity in the intermediate trophic level to only the species that can tolerate large temperature extremes. (iii) Due to the large body size and slow generation time of species in the top trophic level, increased temperature variation is expected to cause high mortality of predators. (iv) In the aquatic community of the *S. purpurea* system and also in larger scale enclosed freshwater systems, they expect the general impact of temperature variation to be stronger in late successional communities compared with early successional communities, as the latter species (fast-growing pioneer species) can adapt more quickly to rapid temperature changes due to faster generation times, than the slower-growing, larger-bodied species found in later succession. (v) In addition, they hypothesize that community respiration will not be markedly affected by increased temperature variation. (Zander et al., 2017)

Lesson 34　Size-balanced community reorganization in response to nutrients and warming

　　Ecosystems consist of many species that interact with their environment and each other and may be depicted as food webs describing 'who eats whom'. The way in which species contribute to the functioning of a food web is constrained by specific biological and environmental conditions, which are subject to alteration by anthropogenic drivers such as the increase in atmospheric CO_2, overexploitation of natural ecosystems and the over application of agricultural fertilizers. In these food webs, 95% of species tend to be no more than three links apart, and seemingly small perturbations to a single species may be swiftly communicated to the entire web. There is potential for such small effects to cause catastrophic phase shifts, for example where the depletion of top predators has led to systems dominated by primary producers or gelatinous consumers such as comb jellies.

　　Due to anthropogenic CO_2 emissions, a global increase in mean surface temperature of at least 1.5–2℃ is expected by 2100 (IPCC 2013). Warming can disrupt the total energy of an ecosystem by directly altering the rate and type of primary production, while simultaneously modifying the ability of consumers to regulate these changes. This strengthening of regulatory top-down effects is in part due to the direct effect that temperature has on accelerating individual metabolism, which varies depending on body size and trophic position. This increased demand for metabolic upkeep leaves less energy available for growth and reproduction, and the avoidance of disease and parasitism.

An indirect effect of future climate scenarios is that differential tolerances to warming between organisms of varying body size are likely to favour smaller-bodied individuals and species, a process that has already been widely observed. Changes in body mass might occur due to the competitive exclusion of larger organisms in favour of smaller individuals and species that need less food, grow faster and reproduce earlier. Many predator populations have already experienced a disproportionate decline in average body size. In aquatic ecosystems, this may also be exacerbated by size-selective harvesting, which leads to the direct removal of the oldest and largest top predators and results in a contemporary evolutionary pressure that selects for rapid growth and early reproduction. Many food webs are characterized by predators eating prey that are within a range of optimal sizes, and a decline in the body size of upper trophic level organisms may result in an allometric trophic cascade, with alternating release from, and increases in, predation pressure across multiple trophic levels. Subsequently, it has become increasingly apparent that changes in size structure have far-reaching effects that are comparable to the removal of entire species. Direct warming and changes in population size structure may also interact to produce synergistic or antagonistic effects that are sometimes seen when other stressors combine. For instance, warming accelerates growth in aquatic systems and fast-growing individuals are more vulnerable to fishing because they forage with greater risk.

Stressors associated with future climate change might also interact with present day stressors where ecological impacts have already been documented. The nutrient status of coastal waters has risen greatly over the last two centuries, a phenomenon that will be exacerbated by climate change-induced increases in the intensity of episodic rainfall events over the coming decades. Eutrophication may promote dominance of a subset of species due to reduced competition for basal resources or lead to alterations in community structure. Nutrient enrichment may increase primary production in some species, but as respiration is more sensitive to warming than photosynthesis, increased temperature might counteract enrichment through increased demand for resources. Additionally, warming might favour smaller-bodied organisms that require fewer resources, but this advantage could be lost when the productivity of a system is enhanced through nutrient enrichment. Thus, the effects of warming and nutrient enrichment might cancel each other out at the community level, although these impacts will be moderated by species-specific responses.

Trophic cascades mediated by global change can shift the flow of energy and nutrients throughout a system, resulting in the creation of new communities which might have altered functioning or productivity. This increases uncertainty for the delivery of economically valuable services that are the higher-level products of fine-scale ecosystem and food web processes. It is not yet known how the effects of direct warming, decreasing predator body size and eutrophication may interact as they are communicated through food webs. Understanding the interplay between the direct and indirect effects of these stressors is critical for predicting ecological responses to the challenges already faced on a daily basis and the likely changes that will be encountered in the near future.

In the experiment performed by Mcelroy et al., a number of hypotheses were tested in temperate mesocosms as mimics of naturally occurring rock pools: (H1) warming increases (a) growth as indicated by moulting rates and (b) mortality of crabs; (H2) warming affects interaction strengths between crabs and the rest of the community; (H3) nutrient enrichment affects interaction strengths between crabs and the rest of the community; (H4) decreasing top-predator body mass affects top-down control with effects cascading to the basal algal assemblage; and (H5) assemblage responses to all three stressors will be interactive, such that the reciprocal effects of decreasing predator body size, warming and nutrient enrichment on producer and primary consumer biomass are likely to influence each other if H2, H3 and H4 are correct. (Mcelroy et al., 2015)

Lesson 35 Functional trait diversity across trophic levels determines herbivore impact on plant community biomass

How biodiversity regulates ecosystem functioning has stimulated much research in the past decade and led to important progress towards answering this question. Pioneering biodiversity experiments have shown a general positive relationship between species diversity and ecosystem functioning. However, it is increasingly recognized that it is not the number of species *per se* that influences ecosystem functioning, but rather the functional traits of species within communities and their diversity. On one hand, the traits of the dominant species are hypothesised to determine biodiversity effects on ecosystem functioning because dominant species represent the largest amount of biomass within communities. On the other hand, functional diversity, i.e. the diversity and the abundance of traits in a community, may affect ecosystem functioning largely because of niche differences between co-occurring species. Increasing niche differences between species is hypothesised to improve resource utilisation and enhance ecosystem functioning.

Most biodiversity experiments conducted in terrestrial systems have manipulated species diversity within one single trophic level, primary producers. Some studies have considered multiple interacting trophic levels and tested how plant species diversity impacts higher trophic levels. In contrast, few studies have explicitly manipulated functional diversity of higher trophic levels. Recent experimental evidence from aquatic systems suggests that functional diversity in higher trophic levels predicts ecosystem functioning with more accuracy than species diversity. Considering the role of trophic interactions for ecosystem functioning, testing the impact of functional diversity within and across multiple trophic levels represents an important step forward to improve our ability to scale up biodiversity effects to ecosystems. To achieve this goal, it would require to explicitly identify the functional traits that mediate trophic interactions between adjacent trophic levels and determine their impact at the ecosystem scale.

Herbivores play a major role for plant diversity and functioning of grassland ecosystems. Yet, studies investigating plant-herbivore interactions at the community and ecosystem scale have mostly focused on single herbivore species or on large groups of vertebrate herbivores. However, herbivore communities, and in particular herbivorous insect communities, can be extremely species rich and functionally diverse, e.g. they exhibit a large variation in body size and feeding niches. How functional diversity within herbivore communities determines their impact on plant communities has never been quantitatively tested.

Grasshoppers represent the largest proportion of arthropod biomass in temperate grasslands, significantly impacting plant biomass and ecosystem functioning. Previous work demonstrated that the impact of single grasshopper species on plant biomass can be precisely predicted from mandibular traits (i.e. incisor strength) because of mechanistic linkages between incisor strength and leaf biomechanical properties. Across grasshopper species, clear feeding niche differences have been reported, with grasshoppers characterised by strong incisor strength consuming tough leaves while grasshoppers with weaker incisor strength preferentially consuming tender leaves. Feeding niche differences between grasshoppers suggest that complementarity may determine herbivore community impact on plant community biomass. Increasing the functional diversity of mandibular traits within a grasshopper community could increase the number of plant species eaten, thus increasing overall plant biomass consumption. However, the complementarity hypothesis only holds if local plant resources are limiting for herbivores. In absence of resource limitation, increasing the functional diversity of mandibular traits would in contrast decrease herbivore impact on plant biomass. At a fixed density of herbivores within communities, an increase in mandibular trait diversity would increase the range of plant species potentially eaten and result in a lower herbivory pressure at the level of plant individuals. This effect is likely to be modulated by plant diversity itself, an important parameter controlling for resource availability.

  Deraison et al. conducted a grasshopper diversity experiment using six functionally contrasting grasshopper species, and manipulated both species richness and functional diversity of grasshopper incisor strength while controlling for variations in grasshopper functional identity. Throughout a 2-year experiment, they sampled plant biomass to assess the impact of grasshoppers. They hypothesised that the functional identity and diversity of grasshopper and plant communities were better predictors of grasshopper impact on plant biomass than species identity and diversity. Specifically, increasing functional diversity of incisor strength would: (1) increase feeding niche differences between grasshoppers at the community level; (2) translate into a higher impact of grasshopper communities on plant community biomass. (Deraison et al., 2015)

Lesson 36 Resource pulses can alleviate the biodiversity-invasion relationship in soil microbial communities

Invasive species have the potential to alter native community structure and affect the functioning of ecosystems. The extent of this impact is tightly linked to the level of biodiversity at a local scale, which plays a key role in buffering a system from invasion. There is no general theory that ties together all the factors influencing what is colloquially known as the diversity-invasion relationship, whereby more diverse communities resist invasion better than less diverse communities. Still, most studies revolve around the role of resource-use mechanisms, whether it be competition between the resident community and the invader or the availability of resources in the environment. Niche theory purports that the probability of an invasion will exponentially decrease as diversity increases due to the increased partitioning of resources among more diverse communities. In line with this theory, a general and emerging principle to consider when investigating biological invasion is that resident communities should resist invasion when their resource uptake best matches resource supply. Should resource supply outpace resource uptake, the chance of invasion should increase. Moreover, the theories proposed by both Davis et al. and Tilman predict that resource pulses will make a community more invasible and facilitate coexistence between an invader and the community. These theories have been supported by empirical examples coming from macroorganisms, and resource pulses have also been shown to further depend on their timing, as well as the life history and uptake capabilities of the invader and resident community members. Interestingly, even though the fluctuating resource hypothesis (FRH) is rooted in the idea that the degree of competition is negatively related to the amount of unused resources, the FRH predicts the absence of any diversity-invasion effect due to the assumption that a resource pulse will suppress competition among the resident and invader populations. While this is logical, never before has this prediction been tested and demonstrated.

Invasion patterns from microbial communities indicate that such communities display diversity-invasion relationships that are similar to those of higher organisms when measured on a local scale. Indeed, reducing the diversity of natural soil communities has shown to increase the survival time of invading *Escherichia coli* and *Listeria monocytogenes*. Experiments creating diversity gradients with simplified, assembled microbial communities have shown similar patterns. For instance, communities containing an increasing number (from one to eight strains) of *Pseudomonas fluorescens* genotypes increasingly resisted invasion from *Pseudomonas putida* and *Serratia liquefaciens*.

Harnessing the capability to manipulate and create microbial communities provides an ideal opportunity to foster our understanding of the mechanisms controlling the diversity-invasion relationship. Microbial assemblage experiments using simplified communities have touched upon the mechanisms that make more diverse communities more resistant to invasion than lower diversity communities. For instance, by assessing the impact of the pathogenic bacterium *Ralstonia solanacearum* on tomato plants in soil microcosms containing five, ten, or fifteen rhizobacterial strains, Irikiin et al. showed that higher frequencies of diseased plants were found in communities of lower diversity and lower resource-use potential. Thus, the large competitive effects in the more diverse communities could have limited the pathogen's growth, survival, and infection capability. In another experiment, van Elsas et al. showed that the competitive effects of richness limited the short-term impact of the invader on community functioning, measured as their ability to use carbon substrates. Moreover, the importance of the community's functional diversity among bacterial communities in conjunction with the number of resources available has highlighted the importance of resource complementarity in invasion resistance. Taken together, these studies encompass the most basic mechanistic understanding of the diversity-invasion relationship. In short, more diverse communities better exploit resources; this stems the invader's access to vital resources and results in its eventual death. However, this mechanism has not been explored throughout more realistic levels of bacterial species diversity. Moreover, from an invader point of view, never before has its actual resource availability been quantified across a diversity gradient, nor have the effects of a resource pulse been examined to understand the extent to which it could permit invasion across a diversity gradient.

Mallon et al. focus on a three-step argumentation to set forth the extent to which resource availability influences the diversity-invasibility relationship for soil microbial communities. Their arguments rely on the assumptions that (1) resource competition suppresses invasion, (2) diversity increases resource competition due to niche preemption, and (3) resource pulses can suppress competition and therefore promote invasibility even in highly diverse communities. Specifically, they test the hypothesis that higher levels of resident microbial diversity reduce the niche available to an invader via increased competitive interactions and that this relationship is alleviated by the application of a resource pulse. (Mallon et al., 2015)

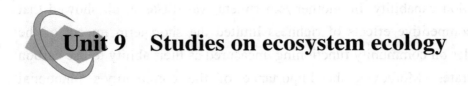

Unit 9 Studies on ecosystem ecology

Lesson 37 Energy flows in ecosystems—relationships between predator and prey biomass are remarkably similar in different ecosystems

All organisms in an ecosystem can be placed on a trophic level, depending on whether they are producers or consumers of energy within the food chain. Ecologists have long debated what regulates the trophic structure and dynamics of ecosystems. This is important because trophic structure and dynamics regulate many of the goods and services that ecosystems provide to wildlife and humankind, such as the production of harvestable food and energy, carbon sequestration and modulation of climate change, and nutrient uptake and control of global biogeochemical cycles. A recent report by Lafferty et al. represents important advances toward a unified theory of trophic structure that captures observed trends across all ecosystems. The ratio of predator-to-prey biomass is a key element of trophic structure that has been studied intensively given its importance for understanding biomass distributions and energy cycling in ecosystems. The nature and control of this ratio have been controversial, but a growing body of literature shows the ratio to be more bottom-heavy in ecosystems with higher prey biomass. In other words, as prey biomass increases, the ratio of predator-to-prey biomass decreases in ecosystems. This pattern has, however, only been demonstrated for specific types of

Unit 9   Studies on ecosystem ecology

ecosystems, such as planktonic systems, and its generality has remained uncertain. Hatton et al. show that this pattern—that is, a decreasing predator-to-prey biomass ratio with increasing prey biomass—applies universally in both aquatic and terrestrial ecosystems. Furthermore, they demonstrate that this universal pattern emerges from a sublinear scaling ($k = 0.75$) that is independent of the ecosystem considered.

Where does this sublinear pattern stem from? Prior research has shown that, in aquatic and terrestrial ecosystems, consumer biomass is linearly related to the consumption of basal resource, which in turn is linearly related to the productivity of the basal resource. Thus, predator (consumer) biomass and prey (basal resource) productivity are linearly related, and if predator biomass is sublinearly related to prey biomass, then prey productivity should also be sublinearly related to prey biomass. In a series of elegant calculations with simple trophic models and fits to empirical data, Hatton et al. demonstrate the universality of these processes. Previous studies have found similar trends in some aquatic and terrestrial ecosystems, but Hatton et al. now generalize their occurrence in nature, thereby advancing substantially our understanding of ecosystem trophic dynamics and structure.

The question follows why prey productivity is scaled sublinearly to prey biomass. If we can answer this question, then we would understand the mechanisms underlying the universal trophic cascade processes shown by Hatton et al. The authors consider tenets of the metabolic theory of ecology. According to this theory, metabolic constraints with increasing individual size generate a sublinear scaling between individual growth (biomass production) and size (biomass), with a sublinear scaling coefficient of $k = 0.75$. Thus, if ecosystems with higher prey biomass are also composed of larger-size prey, it follows that metabolic constraints on body size could explain the sublinear scaling between prey productivity and biomass in ecosystems. Recent evidence has shown this to be the case across ecosystem types differing widely in individual prey size. For instance, when comparing phytoplankton communities to seagrass beds, shrublands, and forests, increasing individual prey size explains the sublinear scaling between ecosystem prey productivity and biomass.

However, when Hatton et al. compared communities within the same ecosystem type, they found that individual prey size does not increase with ecosystem prey biomass and, thus, cannot account for the sublinear scaling pattern. Instead, prey density was higher in ecosystems with higher prey biomass. The authors suggest that processes that depend on prey density, such as competition for resources and other negative interactions among prey species, can result in the sublinear scaling between ecosystem prey productivity and biomass. Another interesting idea is that higher ecosystem biomass, regardless of the size of the prey in the ecosystem, could be subject to the same metabolic constraints on individual body size, thereby averting the need to invoke density-dependent processes to explain the sublinear scaling.

The numerous avenues of new and exciting research opened by Hatton et al. are heightened by the results of Lafferty et al. In an impressive compilation, the authors show that all trophic models published to date, including the seminal Lotka-Volterra predator-prey equations, can be unified into a general consumer-resource population model. The general model contains several quantifiable state variables for consumers and their resources. It can thus be adapted to explain diverse trophic dynamics, ranging from classical examples where the consumer is a predator to cases where the consumer is a micropredator, parasitoid, or parasite. Trophic models that may have been regarded as disconnected and exclusive now emerge as variants of the same conceptual framework. Adaptation of the general model to specific models reveals the simplifications and assumptions that are idiosyncratic to each of them. This provides an accurate procedure to evaluate the focus, limitations, and applicability of all trophic models.

Lafferty et al. bring the patterns found by Hatton et al. to a new level of scrutiny, providing a test for whether such patterns are truly universal. The general trophic model of Lafferty et al. indicates that all consumers and their resources follow the same fundamental principles that govern energy transfer and trophic structure in ecosystems. Thus, the sublinear biomass scaling reported by Hatton et al. for predators and their prey could also apply to any other type of consumer and its resource, including micropredators and parasites. Such patterns could in turn invariably emerge from sublinear scaling between resource productivity and biomass. Confirming these hypotheses would mark a major milestone in ecosystem science. (Cebrian, 2015)

Lesson 38 Diversity increases carbon storage and tree productivity in Spanish forests

Under global change the increase in human-mediated modifications of ecosystems could lead to important losses of biodiversity. Reductions in biodiversity may alter the quality and number of ecosystem functions and services provided by terrestrial ecosystems. Thus, biodiversity-ecosystem functioning (BEF) relationships are an important topic in ecology and have been the subject of considerable debate during the last decades. Most studies of BEF relationships have used species richness as a measure of diversity. However, it has recently been shown that functional diversity better connects the underlying mechanisms of the effects of biodiversity on ecosystem functioning. Trait-based approaches are a promising avenue for disentangling the underlying mechanisms of the effects of diversity on productivity.

Two main, not mutually exclusive, mechanisms of the positive effects of diversity on ecosystem functioning have been proposed: the complementarity and the selection effects. The complementarity effect increases ecosystem function through facilitation and niche partitioning, because functionally diverse species assemblages would enhance resource use efficiency and nutrient retention. Some authors have suggested that complementarity effects could be particularly important in low-productivity or harsh environments, where species interactions are less affected by competitive exclusion, but other authors have observed that complementarity effects are similar across different forest biomes. The selection effect (i.e. selection of particular species or functional traits) proposes that high species richness increases the probability of including the most productive species which will become dominant in the community. Thus, selection effects are partially explained by the "mass-ratio hypothesis" stating that levels of ecosystem function are mainly determined by the functional traits of dominant species. Both complementarity and selection effects simultaneously underlie the net effect of biodiversity on ecosystem function.

Most BEF studies have been conducted in experimental grasslands testing the effects of species richness on ecosystem functions such as biomass production and nutrient cycling. Studies conducted in forest systems, either planted or natural, are much more recent and scarce. These studies have mainly been based on observational forest inventory data and measures of species diversity, and although they have highlighted the importance of functional trait approaches, most of them did not explicitly consider functional diversity and the underlying mechanisms of BEF relationships. The only study that, to our knowledge, has quantified the relative importance of complementarity and selection mechanisms in forest ecosystems suggests that both mechanisms could underlie BEF relationships, at least in simulated mesic temperate forests. More research is needed to understand the role of BEF mechanisms in real forest communities differing in species composition, stand origin and environmental conditions along large bioclimatic gradients.

In this study, Ruiz-Benito et al. conducted a large-scale assessment of two ecosystem functions (carbon storage and tree productivity) along wide climatic, forest structure and diversity gradients using $c.$ 54,000 plots distributed over forests in continental Spain. Continental Spain harbours a large variety of forest types, ranging from Atlantic deciduous broadleaved forests to sclerophyllous and Mediterranean pine forests. Changes in tree carbon storage in Spanish forests depend on climatic and structural conditions, and positive effects of species richness on tree productivity have already been reported. However, the underlying mechanisms and the role of functional diversity on both carbon storage and tree productivity are still poorly explored. Their main objectives are: (1) to analyse the sign and magnitude of BEF relationships in Spanish forests, expecting an effect of diversity on both carbon storage and tree productivity, even when controlling for climatic and structural effects in different forest types; and (2) to understand how complementarity and selection mechanisms affect carbon storage and tree productivity in different Spanish forest types, including natural and planted pine forests. Increasing our understanding of the underlying mechanisms of the effects of diversity on carbon storage and tree productivity is critical for guiding conservation actions and counteracting the effects of species loss on the functioning of forest ecosystems.（Ruiz-Benito et al., 2014）

Lesson 39   Biodiversity effects on ecosystem functioning change along environmental stress gradients

Understanding the influence of biodiversity for ecosystem functioning is crucial as we face extinction rates several orders of magnitude higher than those inferred for the last tens of millions of years, and expected to rise further as a result of climate change, landscape conversion, and other anthropogenic changes. A large body of empirical studies conducted under more or less constant environmental conditions has documented in most cases positive effects of biodiversity on ecosystem functioning. Ecological theory predicts that biodiversity stabilizes and enhances ecosystem functioning, although ideas about this topic have changed over time. Following new experimental results, the current view again is that biodiversity generally buffers ecosystems against stress.

Ecosystems are subject to natural temporal and spatial variation of environmental conditions such as temperature, precipitation and nutrient availability, as well as to influences determined by other species (e.g. predators, competitors, invaders) and human activities. These fluctuations vary in their frequencies and intensities, ranging from limited, regularly recurring variations to which organisms living in a given environment are more or less adapted, to episodic, catastrophic disturbances that lead to extensive mortality and local extinction. If these fluctuations are detrimental to a species or ecosystem function, they are often called stress. One of the major challenges in exploring the impact of stress intensity on biodiversity-ecosystem functioning relationships is that the term stress is a meta-concept that is difficult to define in a general way because a set of conditions that is detrimental (stressful) to one species may be beneficial for another. To circumvent this problem in this study, they use a scheme in which they define the terms stress, stress-response intensity, biodiversity effect and stress-response buffering effect, and they propose an approach to quantify these variables in the experiment performed by Steudel et al.

Empirical studies of biodiversity-ecosystem functioning relationships under stressful conditions are rare and have resulted in contrasting results, ranging from clearly positive to no or in some circumstances even negative effects. This currently unexplained variability of results led to conclude "that current theory (on biodiversity-ecosystem functioning relationships) may still be missing some significant elements". One such previously largely ignored element, both from an empirical and theoretical point of view, is the intensity of stress.

Microalgae have been shown in previous experiments to partly show a positive biodiversity effect on ecosystem functioning. Importantly, these effects were dependent on the environmental conditions prevailing during the experiment and included a negative effect of biodiversity on algal biomass. This suggests that microalgae are suitable organisms to explore the relationship between biodiversity effects and environmental stress. Microalgae show fewer differences in morphological traits than vascular plants, but the diversity of functional or physiological traits as well as of resource utilisation may be similar to those of higher plants.

In the experiment performed by Steudel et al., They set out to explore effects of stress by using both an experimental approach in which they varied stress intensity, and a comparative analysis of published studies assessing the effect of biodiversity-ecosystem functioning relationships under conditions of environmental stress. Their basic hypothesis was that stress intensity affects the strength of biodiversity-ecosystem functioning relationships, although there were no a priory reasons to decide whether a positive or negative net effect of stress intensity on the relationship was to be expected. (Steudel et al., 2012)

Lesson 40 Disturbances catalyze the adaptation of forest ecosystems to changing climate conditions

For long-lived organisms such as trees, the rapid progress of anthropogenic climate change means that they will experience a distinctly different environment toward the end of their life compared to the conditions under which they have established, resulting in disequilibrium between the vegetation composition and the environment. Such a growing maladaptation of the prevailing vegetation to climate is likely to negatively affect the provisioning of a wide range of ecosystem services to society. Furthermore, the occurrence of forest-dwelling species is strongly linked to the prevalence of specific tree species. Increasingly, maladapted forests may thus commit ecosystems to an extinction debt (i.e., a delayed extinction of species due to a protracted response of the ecosystem) and mask the rate and severity of the ongoing biodiversity loss due to a delayed response of tree species to a changing climate. Consequently, rapid climate change induces high uncertainty into the management of forest ecosystems for the provisioning of ecosystem services and the conservation of biodiversity, as experiences made under relatively constant climatic conditions (with climate and vegetation in equilibrium) are increasingly rendered inapplicable.

Theory suggests that disturbance catalyzes change in ecosystems, and can thus reduce the disequilibrium between the prevailing species composition and changing environmental conditions. This notion applies to both natural disturbances in unmanaged systems and silvicultural

Unit 9 Studies on ecosystem ecology

interventions in managed systems, given that management allows adaptation processes such as natural regeneration to ensue after a disturbance. Here, referring to a reduction of the disequilibrium between vegetation composition and climate through natural processes as "autonomous adaptation" (in short referred to as adaptation in the remainder of the text). Processes through which disturbance fosters adaptation include the modification of competition among species, increased resource availability, and a reset of system-level connectedness after disturbance (i.e., a shift from primarily system-internal control mechanisms such as the competition for light to mainly external controls from, for instance, climate and the availability of seeds). As a result of these processes, disturbances can initiate ecosystem reorganization by providing opportunities for new species to invade a site, or giving already present but suppressed species the chance to attain dominance.

Studies that have investigated disturbance-climate relationships generally suggest an intensification of natural disturbance activity in the future as a result of climate change. Hitherto, these disturbance changes have been mainly discussed as a potential threat to ecological resilience. However, based on theoretical understanding of ecosystem dynamics such changes can also be hypothesized to facilitate the adaptation of forests to climatic changes, as more disturbance results in a larger share of landscapes being in the state of reorganization that follows after disturbance. The increasing level of natural disturbance observed in many ecosystems around the globe currently could thus also be interpreted as a mechanism through which ecosystems reduce a growing maladaptation to a rapidly changing climate. The role of disturbance in shaping future ecosystem composition and reducing the climate-vegetation disequilibrium has as of yet been widely overlooked in the discussion of changing disturbance regimes under climate change.

Yet, changes in key ecosystem processes such as disturbances can also lead to profoundly altered ecosystem dynamics in both natural and managed forests. This has the potential to result in ecological novelty and the emergence of no-analog combinations of species. Whether such novel trajectories of ecosystems are compatible with goals of conservation and ecosystem services provisioning remain unclear to date.

Understanding the potential trajectories of forest ecosystems under climate change is thus of paramount importance for ecosystem management. A general proposition frequently found in the ecosystem management literature is that species will shift to higher latitudes and elevations due to global warming. Such trajectories are confirmed by many studies using empirically calibrated species distribution models. Species distribution models are powerful tools representing the fundamental niche of a species, and allowing the potential future distribution of a species' niche to be mapped in relation to projected climatic changes. However, they do assume climate-vegetation equilibrium of the current vegetation, and do not consider relevant processes such as migration and competition among species. As a result, important aspects of adaptation such as time lags in the turnover of the current species composition to a changing future climate are disregarded. Furthermore, as the effects of changes in species interactions in response to a changing climate are not considered, shifts in the realized niche of species and the potential rise of novel species communities remain scarcely investigated.

In the experiment performed by Thom et al., they studied the interactions between vegetation, disturbance, and climate in a complex, unmanaged mountain forest landscape. Their objectives were to (i) assess the time lags in the response of tree species composition and association to changing climatic conditions, (ii) examine the role of disturbance as possible facilitator of this adaptation process, and (iii) study how species turnover in response to different climate and disturbance regimes differs in space. They hypothesized long time lags in the autonomous adaption of the tree species composition to climate change. Furthermore, they expected the adaptation processes to be catalyzed by intensifying disturbance activity, consequently reducing vegetation-climate disequilibrium. Specifically, they hypothesized a stronger influence of disturbance frequency and size than disturbance severity. Finally, they hypothesized that changing climate and disturbance regimes create local novelty in forest ecosystems, that is, future tree species compositions and associations that are currently not present at a site, as both processes alter local environmental conditions, the availability of resources, and the competitive relations between species. In particular, they expected novelty to be most distinctive at low elevations where no-analog environmental conditions will emerge under climate change. (Thom et al., 2017)

Lesson 41　Elevated CO_2 and temperature increase soil C losses from a soybean-maize ecosystem

Human activity, primarily fossil fuel burning, is increasing atmospheric CO_2 and raising global mean temperature. These changes are likely to have direct and indirect effects on storage of soil organic carbon (SOC), but estimates of the direction and magnitude of these effects are poorly constrained. Soils worldwide store over two orders of magnitude more C than annual anthropogenic emissions ($\approx$1,500 Pg C in the top 1 m), so even small changes in soil C storage in response to climate change could produce large feedbacks to the global C cycle. This may be especially true of the SOC-rich former prairie soils of the agriculturally managed Midwestern United States, where annual tillage, infrequent water limitation, regular fertilization, and frequent pulses of highly labile C from crop residues provide ideal conditions for temperature-controlled microbial activity.

Changes in soil C are difficult to detect on short timescales because some pools turn over slowly, with mean residence times of hundreds of years. Although it is conceptually useful to identify the faster-cycling subpools of soil C, we lack experimental methods to measure them directly. Instead, changes in the rate of CO_2 fluxes from soil can be used as a proxy for changes in the soil C cycle by partitioning total CO_2 flux (R_{tot}) into components attributed to "autotrophic" respiration (R_{aut}) from plant roots and rhizosphere organisms, or to "heterotrophic" respiration (R_{het}) from soil microbes in the process of breaking down soil organic matter (SOM). Because R_{het} is the primary avenue for loss of soil C, any change in R_{het} indicates a change in the rate of soil C loss. R_{het} is strongly controlled by soil temperature and moisture and therefore expected to shift under future climate conditions. In contrast, changes in R_{aut} are indirectly linked to the rate of C *input* from roots, so a unit change in R_{tot} could indicate either increasing or decreasing soil C. Therefore, correct partitioning of fluxes is essential to their use as a proxy for changes in pool size.

Previous soil heating experiments have generally shown short-term increases in R_{tot}, except when heating exacerbated soil water limitations. This heating effect often diminishes after a few years of treatment. Whether these responses will persist over the long term under climate change depends on whether a particular soil's R_{het} response is modulated by availability of nutrients or C substrates or by physiological adaptation of the microbial community. In addition, few of these studies were able to separate soil respiration into its autotrophic and heterotrophic components. As R_{het} is strongly controlled by thermal kinetics while R_{aut} responds to a wide variety of nonthermal factors, it has been widely assumed that temperature-associated increases in R_{tot} are driven by increasing R_{het}, but support for this assumption is equivocal.

Previous CO_2 enrichment experiments have generally shown sustained increases in R_{tot}, but there are few reported results from field experiments that manipulate both heat and CO_2 simultaneously. Of those that are reported, the observed responses seem to be mostly mediated by water availability, with heat increasing R_{tot} when moisture is available and reducing it when heating produces drier soil. Elevated CO_2 mediates these effects by ameliorating soil water stress through increased plant water use efficiency, but the strength and predictability of this effect seem to vary widely both within and between experiments.

To measure the root- and SOM-derived components of soil respiration in an intact maize-soybean ecosystem subjected to mid-21st century temperature (+3.5℃) and CO_2 (585 ppm) conditions under fully open-air conditions at SoyFACE (Urbana IL, USA). Black et al. then used a process-based biogeochemical model to predict the long-term effects of these respiratory responses on soil C storage. They predicted that elevated temperature would increase the activity of soil heterotrophs, leading to increased respiration in root-free soil and long-term losses of C from the most labile pools of SOM. They further predicted that elevated CO_2 would increase plant biomass above- and belowground, leading to higher C inputs that would at least partially ameliorate the long-term effect of heat on soil C, and therefore that the long-term fate of soil C at our site would depend on the strength of the interaction between heat and CO_2 effects. (Black et al., 2017)

Unit 10 Studies on global change ecology

Lesson 42 Mycorrhizal status helps explain invasion success of alien plant species

With increasing concern about the effects of invasive alien plants on native plant species, communities and ecosystems, as well as the economic consequences of plant invasion, there has been growing interest in studying the processes and mechanisms underlying successful invasion, including the role of species traits. Besides a set of functional plant traits known to promote invasion, an interest in mutualistic interactions and their influence on invasion success has emerged. Next to pollination, the mycorrhizal symbiosis is the mutualistic interaction that attracted the attention of invasion ecologists. Nevertheless, analyses using mycorrhizal traits to characterize plant species are still rare. Whereas experimental studies, which are mostly local in scale, report the majority of alien plant species to be mycorrhizal, studies based on greater numbers of plant species report ambiguous results. Fitter found that alien plant species were more likely to belong to families that typically associate with mycorrhizal fungi, compared to the native flora of Great Britain. Pringle et al. reported an opposite pattern for alien plant species in California. Hempel et al. showed that neophyte plant species of Germany are more frequently obligate mycorrhizal compared to archaeophytes (introduced before the year 1500) and native species. Therefore, it is still debated whether alien plant species benefit from being mycorrhizal, or whether engaging in the symbiosis constrains their establishment and geographical spread in the new environment and region. A few case studies report positive impacts of the mycorrhizal symbiosis on the growth and development of alien plant species, resulting in a competitive advantage over native species. In a meta-analysis conducted by Bunn et al., the authors did not find a positive correlation between arbuscular mycorrhizal (AM) colonization and growth response in invasive plants, but invasives were more colonized by mycorrhizal fungi, when grown in direct competition with natives. Additionally, the spread of alien plants may be inhibited if required specific fungal partners are not co-introduced.

Relationships with mycorrhizal fungi are of great importance in shaping the ecology of plant species and communities, including those invaded by alien species. Incorporating plant mycorrhizal status and other mycorrhiza-related plant functional traits may thus help to provide further understanding of the establishment of alien plant species and their invasion success. Three groups of plant species can be distinguished according to their mycorrhizal status: (1) obligate mycorrhizal (OM) plant species that are always colonized by mycorrhizal fungi, (2) facultative mycorrhizal (FM) plant species that are colonized under some conditions but not others, and (3) non-mycorrhizal (NM) plant species that are never found to be colonized by mycorrhizal fungi. It is important to note that plant mycorrhizal status and plant mycorrhizal dependency (or responsiveness) are distinct plant traits, not to be confused. While mycorrhizal dependency depicts plant species growth responses under given conditions, mycorrhizal status does not give direct information about the functional significance of mycorrhizal colonization for plant individuals. It rather refers to the mere presence/absence of fungal colonization, and can be used as a proxy for estimating the potential importance of mycorrhizal symbiosis for plants at species level.

The mycorrhizal symbiosis potentially affects the nutrient uptake and C economy of plant species. Depending on mycorrhizal type, mycorrhizal fungi can supply up to 90% of plant P uptake as well as a significant amount of plant N uptake, and can consume up to 50% of a plant's net primary production. Therefore, it is expected about trade-offs between mycorrhizal status and the expression of other plant traits, which require further plant investment, such as the development of morphological structures for storage, dispersal, or vegetative or sexual propagation. Küster et al. demonstrated that trait interactions help explain the invasion success of alien plants in Germany. However, these authors did not include mycorrhizal plant traits. In the experiment performed by Menzel et al., they test for interactions between mycorrhizal status and other functional traits on neophyte invasion success in order to improve their understanding of potential ecological strategies involving the symbiosis.

Although Hempel et al. found that neophytes in the flora of Germany are more frequently OM in comparison with archaeophyte and native plant species, this cannot be used to make inferences about the role of mycorrhizal status in invasion success, as the importance of being mycorrhizal may change during the different stages of invasion. In the study examined by Menzel et al., they aim to answer the following questions: (1) Does the relative frequency of different mycorrhizal statuses (OM, FM, NM) differ between groups of neophyte plant species at different stages of invasion in the German flora, i.e., (a) casual (non-naturalized) species, (b) species naturalized only in human-made habitats, and (c) species also naturalized in habitats with (semi) natural vegetation? (2) Do these groups of neophytes differ from archaeophyte and native plant species in the relative frequency of different mycorrhizal status categories? (3) Do certain combinations of mycorrhizal status and other functional plant traits underlie invasion success? (Menzel et al., 2017)

Lesson 43 Genetically informed ecological niche models improve climate change predictions

Climate change is predicted to be a leading cause of species extinctions in the 21st century and as an agent of selection, is expected to have major impacts on species distributions. One consistent prediction is that changes along established temperature gradients will lead to populations that are either locally adapted or locally maladapted. The ability to respond to these changing gradients depends not only on the rate of environmental change, but also on intraspecific genetic variability, which is critical for an evolutionary response to climate.

Although ecological niche modeling (ENM) has been widely used to predict species' distributions, one aspect of traditional ENMs is that they assume genetic uniformity (i.e., no variation or population structure) throughout a species' range and ignore the potential for local adaptation to specific biotic and abiotic conditions. While relatively few studies have incorporated genetics into niche modeling, to our knowledge no ENM studies to date have explicitly assessed whether incorporating information on genetic variation and local adaptation improves model performance. Simulation studies suggest that incorporating information on genetic variability and population structure can improve our ability to predict species distributions under climate change. Together, these studies underscore the hypothesis that species that are locally adapted to current environments (especially those with limited dispersal capabilities) and in areas with high rates of change are likely to become maladapted in the future. For example, since 1895, the frost-free growing season in the region occupied by *Populus fremontii* has increased as much as 50 days, whereas in the eastern United States at the same latitude, the growing season has changed little. Such variations in both population response and environmental change may greatly affect our ability to predict species distributions using traditional ENMs.

An important consideration for species distribution and ecological niche modeling is whether land management agencies will be able to use this information to plan for the future. Hence, understanding the degree to which current, locally derived stock is matched or mismatched to future environmental conditions is of paramount importance. In fact, forest geneticists are making planting recommendations based upon local stock that is considered to be genetically appropriate for future environmental conditions. The difficulty of making these decisions has prompted recent calls to incorporate both ecological and population genetic data, based on neutral molecular markers, and functional traits, to further inform ecological niche and species distribution models. This recognition is based on the fact that local adaptation allows species to occupy a wide range of environmental niches, especially ones with broad geographic distributions.

To evaluate the utility of incorporating genetic information into species distribution models, Ikeda et al. focused their ecological niche modeling efforts on Fremont cottonwood (*P. fremontii*), a foundation species that is broadly distributed across riparian environments in the southwestern United States and is well known for its ability to both structure communities and influence ecosystem processes. A recent study by Cushman et al. found that populations within the species exhibit significant genetic structure based on neutral molecular markers that defined three genetically distinct clusters. Here, they refer to these clusters as distinct populations or "ecotypes", likely resulting from adaptation to local environmental conditions that differ both in geography and climate among the clusters. Together, these ecotypes encompass the majority of Fremont cotton-wood's distribution in the southwestern United States and occupy specific regions. Hereafter, individuals within these regions will be referred to as the Central California Valley, Sonoran Desert and Utah High Plateaus ecotypes. Additionally, several studies spanning the majority of the range encompassed by these ecotypes have demonstrated that populations within the distribution of Fremont cottonwood are locally adapted to temperature. Specifically, ecotypes varied in leaf phenology such that trees leafing out too early experienced severe frost damage at high elevations, while those that dropped their leaves too late at low elevations suffered a loss in growth, and productivity. These and other mismatches at a given site suggest there is strong selection pressure for local adaptation Thus, the wealth of genetics-based studies with this foundation tree makes it an ideal system for contrasting models with and without population genetic data and to develop the next generation of models that explicitly incorporate genetic information to forecast future distributions. (Ikeda et al., 2017)

Lesson 44   Nitrogen deposition and greenhouse gas emissions from grasslands: uncertainties and future directions

Increases in atmospheric nitrogen deposition (N_{dep}) from human activities can strongly affect the exchange of greenhouse gases (GHG; CO_2, CH_4, and N_2O) between terrestrial ecosystems and the atmosphere. Since the Industrial Revolution, increased N_{dep} (from 17.4 Tg N yr^{-1} in late 1860s to 60 Tg N yr^{-1} in the 1990s) has enhanced terrestrial net ecosystem productivity (NEP) globally by about 175 Pg C. However, corresponding increases in N_2O and CH_4 emissions in response to increased N_{dep} could potentially offset the effect of greater terrestrial carbon storage on the atmosphere. Thus, characterizing the response of GHG fluxes to N_{dep} and determining the thresholds above which the net GHG sink strength declines as N increases are crucial for predicting how terrestrial ecosystems will feedback to climate.

Emission scenarios for the main gaseous forcing agents causing climate change over the 21st century, referred to as representative concentration pathways (RCPs), differ from previous scenarios (e.g., Special Report on Emissions Scenarios, SRES) in that they include regulation of air pollutants such as nitrogen oxides that cause N_{dep}. The RCP predicts that over the coming decades, N_{dep} will increase in most regions of the world. However, the predicted emission of reactive nitrogen to the atmosphere and corresponding rates of N_{dep} are lower in the RCPs compared to SRES scenarios because they include climate policies to reduce their emissions.

Responses of GHG fluxes to N addition are nonlinear and characterized by biological thresholds. Under N limitation, N addition will increase the net GHG sink capacity of ecosystems. At this limiting stage, additional N increases NEP more than it stimulates losses of N_2O and CH_4. As N continues to increase, ecosystems transition into the N saturation-decline stage, and after a critical threshold, their net GHG sink strength will likely decrease. During the intermediate and saturation-decline stages, the sensitivity of NEP to N additions declines at the same time as the emissions of N_2O and CH_4 increase. Increased emissions of N_2O and CH_4 could potentially offset the net CO_2 sink capacity of terrestrial ecosystems including grasslands by 53% to 76%.

Grasslands are important determinants of the concentration of GHGs

Unit 10 Studies on global change ecology

in the atmosphere. They contribute about 10% of terrestrial net productivity and store up to 30% of the world's organic C in their soils. Plant productivity in many grasslands is N limited suggesting that further increases in N_{dep} will increase productivity in these ecosystems into the future, leveling off only when the critical threshold of N saturation is reached.

Considerable effort has been made to understand the impact of N_{dep} on ecosystem C and N dynamics, but loads commonly used in experimental N additions may have been too high to realistically predict the net effect on GHG emissions. Liu & Greaver concluded that most studies of GHG fluxes in grassland ecosystems applied high N loads and very few used multiple N loads. Data on GHG fluxes at low experimental N loads are necessary to accurately predict the net GHG sink strength of grasslands, especially if the responses of GHG fluxes to N_{dep} are nonlinear. It remains unclear if experimental N loads used by most studies are sufficiently low to mimic the loads these ecosystems are predicted to experience in the future. The rarity of GHG data at N_{dep} rates consistent with RCP projections is likely limiting our mechanistic understanding of the relationship between GHG fluxes and N_{dep}, rendering highly uncertain predictions of the net GHG sink strength of grasslands in the future.

Accurate estimates of the net GHG sink strength of grasslands in response to future N_{dep} rates depend on a robust characterization of the relationship between GHG fluxes and N loads, and on the identification of the critical N saturation-decline thresholds. To predict how grasslands will perform in the future, it is important that N loads applied in experiments compare favorably to future rates of N_{dep} projected by RCPs. This is particularly important if grasslands during the 21st century were at the N limitation stage transitioning to the saturation-decline stage.

Many studies predicting the net GHG sink strength of grasslands during the 21st century have assumed a linear relationship between GHG fluxes and N_{dep}. Because the relationship between GHG fluxes and N_{dep} rates is nonlinear, it is likely that predictions of the net GHG strength assuming linearity are inaccurate. Compared to approaches using a nonlinear relationship, predictions of their net GHG sink strength based on a linear relationship would likely underestimate their net GHG sink capacity if grasslands were at N limitation and intermediate stages during the 21st century and would likely overestimate their net GHG sink strength if grasslands were at the N saturation-decline stage.

In this study performed by Gomez-Casanovas et al., to determine whether experimental applications of N and corresponding GHG exchange from grasslands are consistent with RCP projections for future rates of N_{dep} and to predict the relationship between GHG emissions and N_{dep}. Using the process-based biogeochemical model DayCent, parameterized with data from grassland ecosystems located in the major climate zones, they characterize the relationship between N and GHGs fluxes, other N-trace gases fluxes, and N leaching of these grasslands. They used this biogeochemical model to investigate the shapes of the responses of GHG fluxes to changes in N_{dep} (e.g., linear vs. nonlinear) and to determine whether grasslands located in the major climate zones will be at the N limitation and intermediate stages, or at the saturation-decline stage over this century. DayCent has been extensively used to simulate GHG fluxes under different environmental scenarios such as changes in N addition in natural and managed grasslands, providing accurate predictions of ecosystem fluxes of CO_2, N_2O, and CH_4. In addition, DayCent, as other mechanistic-based models, is able to capture the nonlinearity of ecosystem responses to N addition. （Gomez-Casanovas et al., 2016）

Lesson 45　Climate change is not a major driver of shifts in the geographical distributions of North American birds

It is commonly proposed that species geographical distributions are shaped primarily by climate, as follows. First, the set of climatic conditions that a species can tolerate is assumed to be constant, or to change only very slowly. Second, it is assumed that species ranges shift to fill climatically suitable areas. From these assumptions it follows that, as climate changes, species geographical distributions should track climatic change, although potentially with a lag. As Melles et al. put it, "Climate change is thought to be one of the most influential drivers of range shifts".

The evidence regarding this conceptual model is surprisingly mixed. Over the millennia since the Last Glacial Maximum, species distributions shifted in response to changing climate. Yet species did not migrate in unison with climate change. Novel species assemblages formed, and species realized niches changed through time. Changing climate generally accounted for less than half of the variance in the changing composition of species assemblages since the Last Glacial Maximum. Apparently, not all species tracked changing climate closely.

In recent decades, many species' range limits have shifted toward the poles and/or toward higher elevation, apparently in response to climate change. However, meta-analyses have consistently shown that many species' range limits shifted in the opposite direction, or showed little change. Generally, these meta-analyses assumed that climate warmed for all species in the analysis (which need not have been so), and that any poleward or uphill range shifts are causally related to that warming. Further, range-shift studies nearly always focus on occupation of sites at the latitudinal extremes of species ranges, as opposed to shifts of entire ranges.

The main mechanism linking geographical distributions to climate is usually thought to be physiological tolerance. However, interspecific variation in the maximum and minimum temperatures observed within species ranges is only weakly correlated with temperature tolerances observed in the laboratory and the correlations are sometimes negative. Ranges may extend into areas with macroclimates that should not be tolerated on the basis of lab measurements, and ranges may fail to fill areas that should be tolerated. Clearly, macroclimatic boundaries of species ranges are not well predicted by physiological tolerance measured in the lab.

Moreover, factors other than climate are often hypothesized to be major determinants of species geographical distributions. Human-caused habitat loss is described in the conservation literature as the main contributing factor to species decline. Other studies emphasize the influence of interspecific interactions, population and landscape-level processes, land-cover change, neutral processes, etc.

The goal of this study is to test predictions arising from the proposition that shifts in species geographical distributions (not simply their latitudinal extremes) since 1979 are primarily due to climatic warming. Identifying the determinants of species range shifts is important in the context of predicting future shifts in distribution as climate continues to change.

In the study performed by Currie and Venne, they define a species' "range" as the geographical area that circumscribes all sites a species has been observed to occupy. They define "potential range" to mean the range plus a 100 km buffer around it, and use "species distribution" to mean the relative occupancy of sites by a species as a function of a habitat variable (e.g. temperature, latitude). Since niche models often identify temperature as the strongest correlate of species ranges, and since most climate change literature focuses on temperature, this study also focuses on changes in species distributions with respect to temperature.

Currie and Venne examined the geographical distributions, and the distributions of temperatures at occupied sites (i.e. the realized temperature niches), of 21 North American passerine bird species between 1979 and 2010, as documented in the North American Breeding Bird Survey (BBS). Over time, the temperature at any BBS route, and the routes occupied by a given species, may change. For each species they therefore asked: (1) Considering the routes occupied in 1979, did temperatures subsequently change? (2) Did the distribution shift with respect to temperature, latitude or longitude? (3) If so, did the range as a whole (versus a range edge) shift in a consistent direction? (4) Did occupancy of routes change differently in different sections of the ranges (e.g. northern versus southern edges)? (5) Given temperature change and range shifts, did the species maintain a constant realized temperature niche? (6) If temperatures did not remain constant, did range shifts reduce changes in the realized temperature niche, relative to not moving? (7) Are range shifts tracking climate, but with a lag? (Currie and Venne., 2017)

Lesson 46 Temperature impacts on deep-sea biodiversity

Correlation between temperature and marine diversity is one of the most pervasive ecological phenomena not only in the present day but also throughout the last 3 million years, and many ecological and evolutionary hypotheses have been proposed to explain the underlying mechanism for this correlation. However, available large-scale diversity patterns in relation to temperature are still limited to several taxa with sufficient records due to the vastness and inaccessibility of the deep sea, as well as costs associated with technologies needed for deep-sea exploration. Moreover, the different hypotheses proposed so far are still largely controversial, thus further testing and investigations are needed. Since present Intergovernmental Panel on Climate Change (IPCC) scenarios indicate that the temperature of most oceanic regions will change rapidly in coming decades, a better understanding of the potential responses to these changes is one of the main priorities in current ecological research.

Palaeoceanographic and oceanographic studies indicate that deep-sea bottom-water temperature can change over various time scales. For example, during the Cenozoic, the deep sea cooled by more than 10℃ over the last 60 million years. During the Late Quaternary glacial/interglacial cycles, deep-sea temperature was about 4℃ cooler in glacials than in interglacials. Even on millennial and centennial time scales, palaeoceanographic records indicate 1 to 2℃ deep-sea temperature changes. Furthermore, recent oceanographic studies showed dynamic temperature changes even over periods as short as decades or a few years. For example, rapid deep-water warming (∼0.1℃ per decade) over the last about 50 years is known in the western Mediterranean. A recent study in this region also reported rapid drops in temperature (∼0.3℃ cooling within a few years) linked to climate-driven episodic events. In the Labrador Sea, deep-water temperature has shown dynamic decadal variation for the last about 60 years at rates of change up to about 0.5℃ per decade. Abrupt changes in deep-water temperature also have been observed in relation to dense shelf water cascading events, which are able to influence physical and biological processes down to bathyal depths.

Deep-sea temperature shows substantial differences, especially among oceans, depths, and water masses. For example, deep-sea temperature is about 1℃ warmer in the North Atlantic Ocean compared to the North Pacific Ocean at abyssal depths and much warmer at bathyal depths. Some marginal seas such as the Mediterranean, the Red and the Sulu seas have extremely high deep-sea temperature (from ~13℃ for the Mediterranean to >20℃ for the Red Sea at 2000 m depth). Some deep waters at high latitudes are very cold, with temperatures close to −2℃ (e.g. Antarctic Bottom Water). Two deep-water masses in the Atlantic Ocean show distinct temperatures: colder Antarctic Bottom Water and warmer North Atlantic Deep Water. Temperature typically decreases with increasing water depth, except for some very warm intermediate water masses (e.g. Mediterranean Overflow Water in the Atlantic Ocean).

Even though the above-mentioned deep-sea temperature changes and differences in space and time are not subtle, bottom-water temperature has been rather neglected as a possible controlling factor of deep-sea diversity because of its relative stability in space and time compared to shallow-marine systems. However, there is increasing evidence for significant temperature-diversity relationships in the deep sea. Moreover, it is likely that deep-sea organisms are sensitive even to small temperature changes because they live under temperature conditions with much less daily and seasonal variation compared to shallow-marine organisms, although taxa that originated in shallow-marine environments and then penetrated into the deep sea may have stronger tolerance to temperature changes compared to deep-sea taxa originating at depth.

In this review, Yasuhara and Danovaro explore and analyse the relationships between deep-sea benthic alpha (local) diversity and temperature reported from the world's oceans; compare present spatial and past temporal deep-sea temperature-biodiversity patterns; and show that the deep-sea temperature-diversity relationship is positive at low temperatures (<~5℃) and negative at high temperatures (>~10–15℃). When considered over a sufficiently broad temperature range, the temperature-diversity relationship appears to be unimodal, although temperature may be important only at relatively high (>~10–15℃) and low (<~5℃) values and may not play major role at intermediate temperatures. Ecological theories consistently suggest a positive temperature-diversity relationship, but climatic impact projections often suggest negative biological consequences of warming. The unimodal relationship may be one of the keys to solving this fundamental controversy in ecological and climate change sciences. (Yasuhara and Danovaro, 2016)

Part Ⅲ　An Introduction to Commonly Used Foreign Language Databases and Writing Skills for SCI Papers
常用外文数据库介绍及 SCI 论文写作技巧

Unit 11　SCI 及常用外文数据库介绍

Lesson 47　SCI 简介

　　SCI（Science Citation Index，科学引文索引）是美国科学情报研究所（Institute for Scientific Information，ISI）出版的一个世界著名的期刊文献检索工具，其通过严格的选刊标准和评估程序来挑选刊源，而且每年略有增减，从而使 SCI 收录的文献能够全面覆盖全世界最重要和最有影响力的研究成果。SCI 收录的内容涵盖自然科学、工程技术和生物医学等 150 多个学科领域。人们通常所说的"SCI 论文"即是被 SCI 期刊收录的学术论文。

　　SCI 出版形式有四种：印刷版、光盘版（SCI CDE）、联机版（SCI Search）和网络版（SCI-Expanded，SCIE）。经过 50 多年的发展完善，SCI 已从开始时单一的印刷型发展成为功能强大的电子化、集成化、网络化的大型多学科、综合性检索系统。随着网络版的普及，经我国科学技术部有关部门和领导研究，决定从 2000 年的统计工作起，SCI 论文统计检索系统改为用 SCIE。下面将主要介绍 SCI 影响因子计算方法及分区标准。

一、SCI 影响因子计算

　　影响因子（Impact Factor，IF）是美国科学情报研究所期刊引证报告（Journal Citation Report，JCR）中的一项数据，指的是某一期刊的文章在特定年份或时期被引用的频率，是衡量学术期刊影响力的一个重要指标。自 1975 年以来，汤森路透集团每年都会出版期刊引用报告（JCR），提供一套统计数据，展示科学期刊被引用情况、发表论文数量以及论文的平均被引用情况。一份期刊前两年中发表的"源刊文本"（sourceitems）在当年度的总被引用数，除以该期刊在前两年所发表的"引用项"（citableitems）文章总篇数，即为该期刊当年度的影响因子数值。用分数式表达为

期刊 N 年度的影响因子

$$=\frac{该刊(N-2)+(N-1)年所有源刊文本在N年度总被引用数}{该刊(N-2)+(N-1)年发表引用项总数}$$

按照 ISI 给出的定义，期刊的所谓"源刊文本"，指的是该 SCI 期刊发表的所有文本，而"源刊文本"又被区分为"引用项"和"非引用项"（uncitable items）两类，在通常情况下，"引用项"对应着学术文本，"非引用项"对应着非学术文本（穆蕴秋和江晓原，2016）。

在同一学科中，影响因子越高，一定程度上表明该期刊中的论文被引用的机会越大，影响力也越大。

二、SCI 分区标准

由于不同学科之间的 SCI 期刊很难进行比较和评价，根据 JCR 分区表对 SCI 论文进行评价的模式已被国内部分高校和科研机构采纳，国内主流参考的 SCI 分区依据主要有中科院 JCR 分区表和汤森路透 JCR 的 Journal Ranking 分区两种。其中，中科院期刊分区表被更多的机构采纳作为科研评价的指标。因此，下面对中科院 JCR 分区表进行详细介绍。

1. 中科院 JCR 分区表对所有期刊的学科划分

大类学科：医学、生物、农林科学、环境科学与生态学、化学、工程技术、数学、物理、地学、地学天文、社会科学、管理科学及综合性期刊，共 13 个大类。

小类学科：即 JCR 学科分类体系 Journal Ranking 确定的 176 个学科领域。

需要注意的是，一本期刊只可属于一个大类学科，但是一本期刊

却可以属于多个不同的小类学科。

中科院期刊分区也是基于一个期刊的影响因子，但是采用的是三年平均影响因子，其优点是减少了一些期刊影响因子逐年波动较大带来的不稳定因素。

2. 中科院 JCR 分区表划分方法

中科院根据 13 个大类学科中 SCI 期刊近三年平均影响因子划分为 1 区（最高区）、2 区、3 区和 4 区四个等级。中科院分区中的 1 区到 4 区的期刊数量不等，呈金字塔状分布。前 5%为该类 1 区、6%～20% 为 2 区、21%～50%为 3 区，其余为 4 区。此外，中科院分区表的大类分区中还会遴选出一些优秀的 Top 期刊，1 区期刊直接划入 Top 范围内，2 区中两年总被引用频次指标位于前 10%的期刊也归入 Top 期刊集合。

Lesson 48　常用外文数据库介绍

外文数据库是大学生、研究生及科研工作者进行科学研究及撰写论文的重要参考文献来源之一。近年来，许多高校都投入大量的经费用于购买外文数据库，然而，调查发现高校大学生及研究生对外文数据库的利用率非常低，甚至不少研究生根本不会利用外文数据库查找文献。下面将对与生物学相关的国外主流数据库及其使用方法进行简单介绍，希望大家能够借助这些工具来提高自己。

1. ScienceDirect 数据库

Elsevier 是荷兰一家全球著名的学术期刊出版商，ScienceDirect 是荷兰 Elsevier 公司出版的全球最全面的全文文献数据库，涵盖了几乎所有学科领域。每年出版大量的学术图书和期刊，大部分期刊被 SCI、SSCI、EI 收录，是世界上公认的高品位学术期刊。ScienceDirect 得到了 70 多个国家的认可，是目前国内使用率最高、下载量最多的科学数据库。该数据库主页界面如图 11.1 所示。

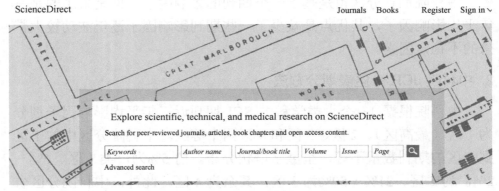

图 11.1　ScienceDirect 数据库主页界面

2. Springer Link 数据库

Springer Link 是德国施普林格出版集团（Springer-Verlag）研制开发的一个在线全文电子期刊数据库。该出版集团以出版图书、期刊、工具书等学术性出版物而著名，通过 Springer Link 系统发行电子图书并提供学术期刊在线服务，为科研人员及科学家提供强有力的信息中心资源平台。当前，Springer Link 数据库收录了 2,700 多种全文期刊，并且所收录的期刊学术价值较高，大部分为 SCI、SSCI 和 EI 收录的核心期刊。该数据库的主页界面如图 11.2 所示。

3. Wiley Online Library 数据库

Wiley Online Library 该数据库是全球历史最悠久、最知名的学术出版商之一，享有世界第一大独立的学协会出版商和第三大学术期刊出版商的地位。2010 年 8 月，Wiley 正式向全球推出了新一代在线资源平台"Wiley Online Library"以取代已使用多年并获得极大成功与美誉的"Wiley InterScience"。同时，所有的内容都已转移至新的平台，确保为用户和订阅者提供无缝集成访问权限。作为全球最大、最全面的经同行评审的科学、技术、医学和学术研究的在线多学科资源平台之一，"Wiley Online Library"覆盖了生命科学、健康科学、自然科学、

Unit 11　SCI 及常用外文数据库介绍

社会与人文科学等全面的学科领域，收录了来自 1,500 余种期刊、10,000 多本在线图书以及数百种多卷册的参考工具书、丛书系列、手册和辞典、实验室指南和数据库的 400 多万篇文章，并提供在线阅读。该数据库的主页界面如图 11.3 所示。

图 11.2　Springer Link 数据库主页界面

图 11.3　Wiley Online Library 数据库主页界面

4. ProQuest 数据库

ProQuest Information and Learning 公司通过 ProQuest 平台提供 60

多个文献数据库，包含文摘题录信息和部分全文。自 2012 年起，原剑桥科学文摘（Cambridge Scientific Abstracts，CSA）平台的数据库全部合并到 ProQuest 平台。这些数据库涉及商业经济、人文社会、医药学、生命科学、水科学与海洋学、环境科学、土木工程、计算机科学、材料科学等广泛领域，包含学位论文、期刊、报纸等多种文献类型，尤其值得一提的是著名商业经济数据库 ABI 和全球最大的学位论文数据库 PQDT，还有原 CSA 平台丰富的特色专业数据库都包含其中。其中 PQDT 是世界上最早及最权威的博硕士论文收藏和供应商，可提供 200 多万篇国外高校博硕士论文的全文，学科覆盖了数学、物理、化学、农业、生物、商业、经济、工程和计算机科学等，是学术研究中十分重要的参考信息源。PQDT 数据库的主页界面如图 11.4 所示。

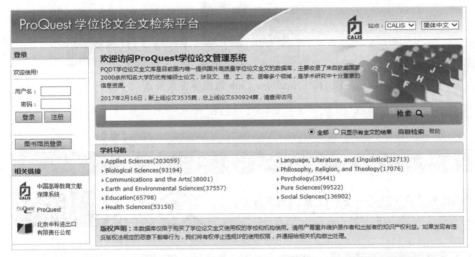

图 11.4　ProQuest 数据库中学位论文全文数据库（PQDT）主页界面

5. NCBI 数据库

美国国立生物技术信息中心（National Center for Biotechnology Information，NCBI），是由美国国立卫生研究院（NIH）于 1988 年创办。创办 NCBI 的初衷是为了给分子生物学家提供一个信息存储和处理的系统。除了建有 GenBank 核酸序列数据库（该数据库的数据资源来自全球几大 DNA 数据库，其中包括日本 DNA 数据库 DDBJ、欧洲分子生物学实验室数据库 EMBL 以及其他几个知名科研机构）之外，NCBI 还可以提供众多功能强大的数据检索与分析工具。目前，NCBI 提供的资源有 Entrez、Entrez Programming Utilities、My NCBI、PubMed、PubMed Central、Entrez Gene、NCBI Taxonomy Browser、BLAST、BLAST Link (BLink)、Electronic PCR 等共计 36 种功能，而且都可以在 NCBI 的主页 www.ncbi.nlm.nih.gov 上找到相应链接，其中多半是由 BLAST 功能发展而来的。该数据库的主页界面如图 11.5 所示。

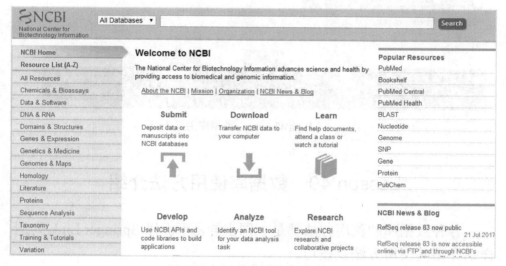

图 11.5　NCBI 数据库主页界面

6. HighWire Press 数据库

　　HighWire Press 是提供免费全文的、全球最大的学术文献出版商之一，于 1995 年由美国斯坦福大学图书馆创立。目前已收录电子期刊 882 多种，文章总数已达 282 多万篇，其中超过 103 万篇文章可免费获得全文，这些数据仍在不断增加。收录的期刊覆盖以下学科：生命科学、医学、物理学、社会科学。该数据库的主页界面如图 11.6 所示。

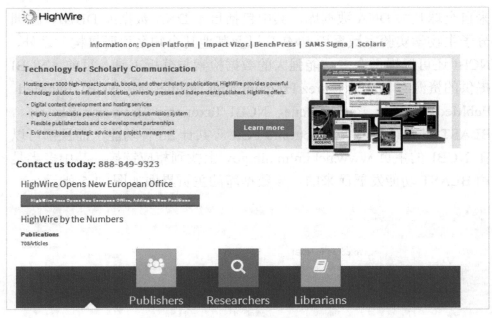

图 11.6　HighWire Press 数据库主页界面

Lesson 49　数据库使用方法介绍

各种数据库的使用方法都是大同小异，下面以 Springer Link 为例详细介绍其使用方法及技巧。

一、简单搜索

在 Springer Link 首页上方有一简单检索框（Search），可直接输入关键词进行全文检索。如图 11.7 所示。

Unit 11　SCI 及常用外文数据库介绍

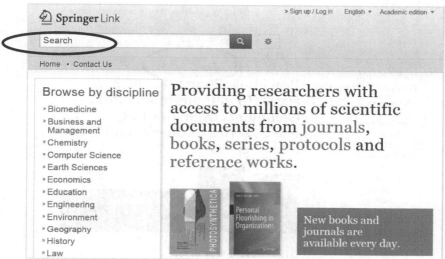

图 11.7　简单检索界面

二、高级检索

在 Springer Link 主页上还提供"高级检索（Advanced Search）"和"检索帮助（Search Help）"，如图 11.8 所示。读者可以通过使用高级搜索选项进一步缩小搜索范围，如图 11.9 所示。例如：检索论文题目中包含"ecology"一词的论文，仅需要在"where the title contains"词条下面输入"ecology"，然后点击图 11.9 下端的"Search"。检索结果页如图 11.10 所示，默认情况下显示所有"Title"中含有"ecology"的结果，在页面左方有聚类选项可以帮助读者优化搜索结果，聚类选项包括：文献内容类型、学科、子学科、语言和出版时间等。

图 11.8　选择高级检索界面

图 11.9　高级检索界面

Unit 11　SCI 及常用外文数据库介绍

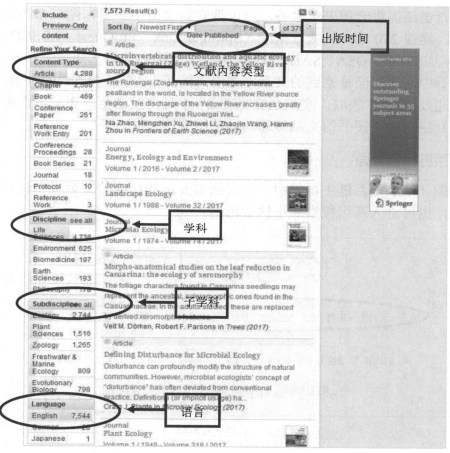

图 11.10　检索结果界面

Lesson 50　百链云服务平台简介

外文全文数据库比较昂贵，一些高校（尤其是地方院校）和研究院所没有订购外文全文数据库，这给高校师生的教学科研工作带来极大的困扰。在此背景下，超星公司推出了新一代图书馆资源解决方案及共建共享方案——百链，它内置丰富的全文资源，其宗旨在于为读者提供资源补缺服务，能够实现本馆与其他馆的互联互通、共建共享，最终通过原文链接和云服务模式，帮助读者找到所需资源。

百链还提供云图书馆服务，可以搜索到全国 700 多家图书馆各类资源，涵盖 270 个中外文数据库（Springer Link、ProQuest、EBSCO、Wiley、JSTOR、OSA、SAE、SIAM 等外文库和中国学术期刊、万方、维普等中文库），当检索到的文献本馆没有购买，不能提供下载的时候，可以通过 E-mail 获取，文献全文将在 24 小时内发送至邮箱。下面将简单介绍百链云图书馆文献传递功能使用方法。

一、百链检索界面

图 11.11 是百链主页界面，可根据需要选择相应的文献类型进行检索，也可进入高级搜索缩小检索范围。

图 11.11　百链主页界面

二、检索结果查看

进入检索界面后可以根据出版时间、学科、文献来源和期刊刊种进一步缩小检索范围（见图 11.12）。

三、查看文章详细信息

找到自己所需要的文献点击其标题，进入文章详细信息，从图 11.13 右侧的"获取途径"中可通过邮箱方式获取全文。

Unit 11　SCI及常用外文数据库介绍

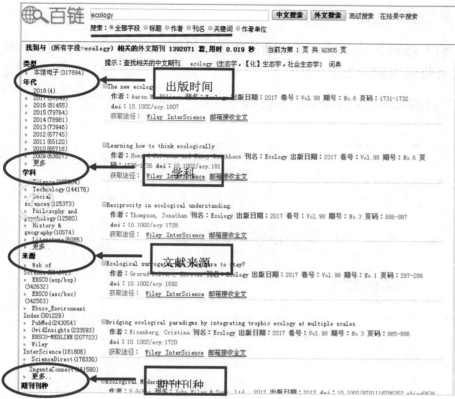

图 11.12　百链检索页界面

图 11.13　文献具体信息界面

四、邮箱获取方式

输入邮箱和验证码，点击确认提交（见图 11.14）。一般在 24 小时内就会收到文献全文。

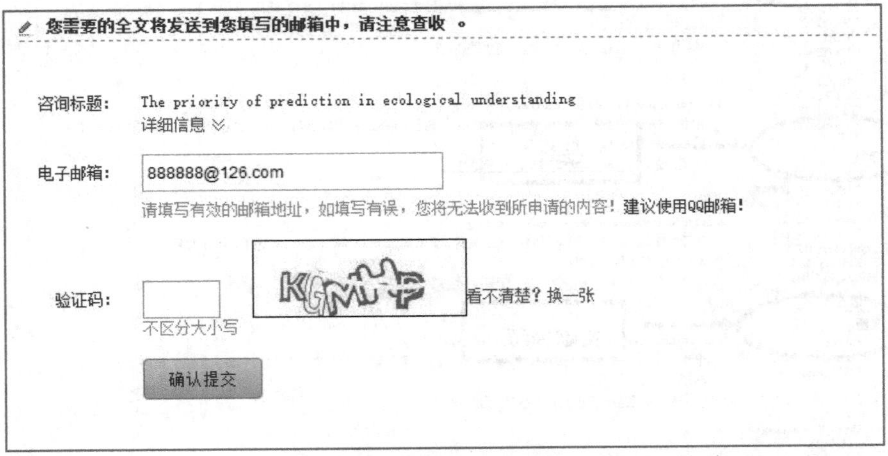

图 11.14　邮箱获取文献方式

Unit 12　SCI 期刊论文写作技巧

科研论文是对科学领域中的问题进行总结、研究、探讨，表述科学研究成果的论文。撰写科研论文是科研工作的最后环节，也是科研成果产出和学术交流的主要形式。可以说，撰写科学论文是一个科研工作者必须具有的基本功。而这个基本功的形成，需要对论文写作方法和技巧进行全面学习，并反复实践才能完成。本单元主要介绍 SCI 期刊论文写作技巧。

Lesson 51　退稿的常见原因

要想写出一篇好的论文，并顺利被编辑部接收和出版，就需要先了解一下论文退稿的常见原因，这样更有利于提高论文投稿的成功率。通常情况下，一篇论文投到编辑部到最后接收发表需要经历以下几个状态（Status）：

Manuscript Submitted ⟶ With Editor ⟶ Under Review ⟶ Required Review completed ⟶ Decision in Process ⟶ Completed Reject/ Reject and Resubmit /Major Revision/Minor Revision ⟶ Accept。

论文要经过一系列的审稿过程，再加上各个期刊的投稿量非常大，审稿过程异常严格。论文在这个过程中有可能在 Manuscript Submitted、With Editor 和 Decision in Process 状态下被退稿。下面将分别介绍论文在这三种状态下被拒的可能原因，以期作者在试验设计和写作过程中注意。

一、论文在"Manuscript Submitted"状态下退稿分析

"Manuscript Submitted"表示论文已成功投递到作者所投期刊，在转给期刊编辑之前，通常会由技术检查人员检查论文格式。如果技术检查符合期刊要求则论文将进入下一环节"With Editor"，如果论文格式不符合期刊要求则会被退稿，要求修改格式后再重新递交。在这一环节论文被退的主要原因可能存在以下几方面。

1. 语言问题

SCI期刊论文使用的是国际通用语言中的英文，因此，对于中国人来说，英文SCI论文的撰写存在一定的难度。文章投稿后，技术检查人员通常会提出论文可读性差，常建议找母语是英文的专家修改后再投。因此，要想克服这一困难，就需要我们有扎实的英语基本功及SCI论文写作技巧。关于SCI论文各部分的写作技巧在后面的章节会详细介绍。

2. 没有行号和页码

一般期刊都要求稿件中插入行号和页码，这样便于审稿专家审稿。行号一般在左侧插入，并且连续编码，页码一般放在每页底部。

3. 参考文献格式不正确

每个杂志都有不同的参考文献格式，因此一定要仔细查看投稿指南，根据要求修改参考文献格式。

4. 图表不能嵌入文中

虽然最终文章发表的印刷版中图表嵌入在文中，但是投稿的稿件不要把图表和文章内容放在一起。通常是将图表放在文章的最后，这种格式审稿人看起来比较舒服。

5. 没有"Competing financial interests"和"Contributions"说明

有些期刊要求对"Competing financial interests"和"Contributions"

进行说明。"Competing financial interests"就是要陈述作者的文章有没有与其他任何单位、机构或任何个人有利益冲突。该陈述一般放在首页末尾或文章最后。如果作者的文章不存在任何利益冲突可表述为：Competing financial interests: The authors declare no competing financial interests.

有些期刊要求"Contributions"说明，即有关作者贡献说明。例如：Author Contributions：Y.J.L., C.Z.Z. and Q.L. conceived the experiments, Y.J.L., D.D.S. and D.D.L. conducted the experiments, Y.J.L. and Z.F.X. analysed the results. Y.J.L. wrote the main manuscript text, and C.Z.Z. revised the manuscript. H.H.L. and Q.L. initiated and supervised the research. All authors reviewed the manuscript.（Li et al. 2015）

二、论文在"With Editor"状态下退稿分析

"With Editor"这个状态表示论文已通过技术审查，编辑开始处理，如果编辑认为作者的论文内容适合期刊，就会送交同行评审，如果论文与期刊的范畴不相符或是没有达到期刊的标准，就有可能在外审前退回给作者，一般会在几天内收到拒绝通知。在这一环节论文被退的主要原因可能存在以下几方面。

1. 论文内容不符合该期刊的范畴

每个期刊的主页上都有该期刊的介绍，因此，投稿前一定要认真查看期刊的"目标和范围（aim and scope）"，以确定稿件内容是否符合该期刊的范畴。

2. 论文水平达不到期刊要求

每种期刊每年出版的论文数量非常有限，所以，多数情况下尽管作者的论文主题在期刊要求的范畴内，同样会被退稿。造成这种退稿的原因主要是论文水平达不到期刊要求，具体体现在论文的创新性、论文中所包含的数据和论文统计方法是否正确等。要避免论文在这一环节被拒，一方面，需要作者在试验设计时认真查阅文献资料，避免

简单地重复别人的工作，尽量提高自己工作的创新性；另一方面，投稿前最好查一下和自己研究相近的论文都在哪些期刊上发表，然后再结合自己论文质量进行选择，这样就可以做到有的放矢。

3. 论文重复率高

编辑部一般会对收到的论文进行查重，如果作者论文与自己先前课题组发表的论文或他人已发论文重复率过高，则同样会被退稿。

三、论文在 "Decision in Process" 状态下退稿分析

一篇论文如果能够通过技术检测（Manuscript Submitted）和编辑审查（With Editor），就会进入专家评审阶段（Under Review），编辑部会找2~3位专家评审，外审专家的意见非常重要，编辑部最后会根据专家意见和主编意见进行综合分析决定论文的结果。其结果主要有以下几种情况。

1. Minor revision/Major revision

如果作者收到稿件处理结果是"Minor revision/Major revision"，只要按照审稿人和编辑部提出的意见认真修改，一般都会接收。

范例：Editorial decision

We have received the reports from our advisors on your manuscript which you submitted to Plant Growth Regulation. Based on the advice received, I feel that your manuscript could be reconsidered for publication should you be prepared to incorporate major revisions. When preparing your revised manuscript, you are asked to carefully consider the reviewer comments which are attached. We look forward to receiving your revised manuscript within eight weeks.

2. Reject with an invitation to resubmit

这种结果要求作者修改后再递交，论文将可能作为新的稿件重新处理。

范例：Editorial decision

Reject with an invitation to resubmit. The manuscript appears to include a significant body of potentially worthy results. Both reviewers have raised a number of significant concerns with regard to the content, organization, format and references in the paper and I therefore do not accept your manuscript for publication in its present form. I would be willing, however, to reconsider a properly rewritten and re-submitted manuscript improved according to all the constructive and extensive suggestions made by the reviewers. A re-submitted manuscript will be subject to a second round of review.

3. Completed Reject

论文被"Completed Reject"的主要原因可能存在以下几个方面：
（1）语言表达有问题，外审专家读不懂。
（2）试验设计不合理。
（3）文章创新性不够。
（4）数据统计或分析有严重缺陷。
（5）结果自相矛盾。
（6）讨论不深刻（描述性讨论）。

总之，SCI 期刊拒稿是一件很正常的事情。因此，作者要保持良好的心态，不要因为被拒而失去信心，而是应该针对论文中出现的问题进行分析和思考。

Lesson 52　SCI 期刊论文写作前的准备及基本框架

一项科研工作基本完成之后，动手写作之前，需要缜密构思，尽力做到条理清晰。因此，要想论文被顺利接收并发表，论文写作前的准备至关重要。本课主要介绍 SCI 期刊论文写作前的准备及论文基本框架。

一、SCI 期刊论文写作前的准备

1. 确定论文主题

研究生期间会做大量的试验，获得大量的数据。因此，写论文前首先要确定论文主题，即作者通过这篇论文想表达什么，揭示什么问题，从而从大批数据中筛选出能够支撑该篇论文的数据。

2. 数据统计分析

加拿大著名生态学家 Pielou 曾经指出"生态学本质上是一门数学"（Pielou，1985）。对于生态学研究者而言，应用统计学方法处理试验数据，获得统计学结论，分析生态学现象，揭示生态学内在规律，是当代生态学研究的常用手段之一。正确应用统计学方法，可以从中发现生态学规律，解释自然生态现象（张峰等，2006）。因此，作者在写论文前一定要先对筛选数据进行统计分析。方差分析和回归分析是生态学中最常用的两种分析方法，常用统计软件为 SPSS（Statistical Package for the Social Science）。SPSS 系统的特点是操作比较方便，统计方法比较齐全，绘制图形、表格方便，输出结果比较直观。

3. 图表制作

在 SCI 论文中，为了清晰地展现处理间差异和试验结果，作者常常会使用图和表表示。表格的优点在于可以方便地列举大量精确数据或资料，而图形则可以直观、有效地表达复杂数据。因此，作者可以根据实际情况选择使用。作者投稿中图表常见问题如下：

（1）表格没用三线表。

（2）表格中同一指标数字或图形中纵坐标轴上的数字有效位数不一致，应保留一致，不足者以"0"补足。

（3）单位不规范，即没有用国际单位。另外，在数字和单位之间应有空格。

（4）无表题或表题不确切或重点不突出。表题是表格和图的重要组成部分，是对表中内容的概括，表题应简洁准确。

(5)图表中数据没有统计分析,没有标出误差和统计学上显著差异的字母标记。

4. 搜集文献

搜集文献的过程就是变相寻找思路的过程,在这个过程当中我们可以查找到适合自己研究的方向并做出一些针对性分析,为作者论文中"引言"和"讨论"的写作提供帮助,同时还有助于目标投稿期刊的选择。

5. 选定目标投稿期刊

对于经验丰富的科研工作者,一般在撰写文章时心中就已经有了安排,而对于大学生和研究生,由于缺乏该方面的经验,通常会觉得无从下手。如何选择期刊是一门学问。选择恰当的期刊是论文能够快速发表的一个重要环节。关于期刊选择的技巧将在后面进行详细介绍。

6. 明确论文的著作权

一项科研成果及 SCI 论文往往由多人或多个单位共同完成。因此,关于论文的署名,作者需要考虑以下几个问题:

(1)通信作者确定:通信作者一般是本项目研究负责人,实际统筹处理试验设计、论文修改、投稿和承担答复审稿意见等工作的主导者,一般都为导师或部门负责人等。

(2)第一作者确定:第一作者是仅次于通信作者的项目主要参与者,贡献多为试验操作、论文撰写等实际执行者。一般为通信作者的学生或课题组成员等。

(3)其他作者的顺序确定:通信作者和第一作者除外的次要参与者和一般参与者,根据贡献大小确定其顺序。

(4)每个论文署名都必须经过作者同意。

关于论文具体署名原则将在"题目、作者及通信地址"一文中进行详细介绍。

二、SCI 论文的基本框架

一篇优秀的 SCI 论文不仅要拥有好的文章内容，并且文章在 SCI 论文结构方面也要清晰明确。SCI 期刊上发表论文的类型主要有：Original Research Papers（Full Length Paper，原创性研究论文），Review Articles（综述）和 Short Communications（简报）。研究生通常以发表"Original Research Papers"为主，下面将以"Original Research Papers"为例介绍 SCI 论文的基本框架。

"Original Research Papers"主要由以下几部分组成：

（1）题目（Title）

（2）作者（Author）

（3）通信地址（Affiliation (s) and address(es)）

（4）摘要（Abstract）

（5）关键词（Keywords）

（6）前言（Introduction）

（7）材料与方法（Materials and methods）

（8）结果（Results）

（9）讨论（Discussion）

（10）结论（Conclusion）

（11）致谢（Acknowledgements）

（12）参考文献（References）

（13）附录（Appendices）

Lesson 53　题目、作者及通信地址撰写技巧

一、题目（Title）

论文题目主要是用来帮助文献追踪、检索和吸引读者，多数期刊对题目有一定的字数限制，因此，应以最少数量的单词来充分表述论文的内容。

1. 题目写作要求

（1）题目应准确、简洁和清楚。

（2）论文题目通常由名词性短语构成，如果出现动词，多为分词或动名词形式。

（3）一般不使用缩略词、化学分子式和专利商标名称等。

（4）题目的长度和字母的大小写应参考投稿期刊中的作者指南及该期刊近期发表的论文。

（5）题目中不使用没有实际意义的词，像 effects of，influence of，characterization of，treatment of，use of，studies of (on)，a few，some 等。

2. 案例分析

原题目：Study on drying methods on the active and nutritional ingredients in postharvest flowers of medicinal chrysanthemum

修改后：Nutritional and Active Ingredients of Medicinal Chrysanthemum Flower Heads Affected by Different Drying Methods

二、作者（Authors）

科技论文署名是一件极其严肃的事情，应按研究工作的实际贡献大小确定署名顺序。每位作者都必须就论文的全部内容向公众负责。国际医学期刊编辑委员会（ICMJE）有关作者资格的界定包括三条：第一，参与课题的构思与设计，资料的分析和解释；第二，参与论文的撰写或对其中重要学术内容做重大修改；第三，参与最后定稿，并同意投稿和出版。必须同时具备这三个条件才能成为作者。

1. 中国人名书写格式

按照欧美国家的习惯，名（first name）在前，姓（surname / family name / last name）在后。中国人在国际期刊上发表文章时人名书写格式主要有以下几种（以张三丰为例）。

Sanfeng Zhang, San-Feng Zhang, Sanfeng ZHANG, San-Feng ZHANG, Zhang Sanfeng, Zhang San-feng。其中最常见的写作格式是，Sanfeng Zhang 和 San-Feng Zhang。

一般国外期刊会尊重作者对自己姓名的表达方式，但前提是一定要告诉编辑部，否则编辑部会默认为放在后面的是姓。不管作者采用哪种写作方式，应该保持一致。

2．署名原则

（1）论文的执笔人或主要撰写者应该是第一作者，例如，一般来讲研究生写的论文应以研究生为第一作者，而不能以导师为第一作者。

（2）不可故意将知名人士署为作者之一，要避免"搭车"现象。

（3）不能故意遗漏具有署名权利的作者。

（4）坚持原则，既不要随便增加作者，也不要随便被增加。

三、通信地址（Corresponding addresses）

通信地址通常包括作者的工作单位、所在城市、邮政编码、国家、邮箱和联系电话等。其中工作单位要参照单位对外正式名称，一般不用缩写。该信息要求准确清楚，其目的一方面有利于不同单位每年成果的统计和考核；另一方面使读者能够按所列信息顺利地与作者联系。

1．通信地址写作要求

（1）工作单位要详细，对于高校而言一般写到院系，例如：The College of Life Sciences, Hebei University。

（2）有两位或两位以上作者，如果作者属于不同单位，则通信地址应按作者的先后顺序列出，并在作者名字上以上标符号的形式标注。例如：San Zhang[a], Si Li[b], Wu Wang[c]

[a] The College of Life Sciences, Hebei University; [b] The College of Life Sciences, Hebei Agriculture University; [c] The College of Life Sciences, Hebei Normal University

（3）如果第一作者不是通信作者，作者应该按期刊的相关规定标出通信作者，多数期刊以星号（*）标注。例如：San Zhang，Si Li，Wu Wang*。

（4）如果第一作者或通信作者涉及多个单位，要根据贡献大小排列顺序。

Lesson 54　摘要和关键词撰写技巧

一、摘要（Abstract）

摘要是 SCI 论文中最核心部分，虽然摘要位于正文前，但摘要通常在正文完成后撰写。摘要主要是为读者阅读、信息检索提供方便。摘要对于初写作者而言比较难，尤其对于研究生来讲，由于大学期间主要以写实验报告为主，因此学生们在写摘要时总是摆脱不了实验报告模式的束缚。其实如果明白摘要的主要构成要素，摘要写作相对来讲比较简单。下面以研究生在写作中常用的"报道性摘要"文体为例介绍摘要的构成要素、写作注意事项及案例分析。

1. 摘要的构成要素

摘要一般由四部分组成：研究目的、研究方法、研究结果和结论，有的期刊要求在研究目的前加上研究背景。

（1）研究目的——准确描述该研究的目的，说明提出问题的缘由，表明研究的范围和重要性。

（2）研究方法——简要说明研究课题的基本设计，结论是如何得到的。

（3）研究结果——简要列出该研究的主要结果，有什么新发现，说明其价值和局限。

（4）结论——简要地说明经验，论证取得的正确观点及理论价值或应用价值，是否还有与此有关的其他问题有待进一步研究，是否可推广应用等。

2. 摘要撰写要求

（1）摘要应具有独立性和自明性，并拥有一次文献同等量的主要信息。因此，摘要是一种可以被引用的完整短文。

（2）摘要写作中时态可大致遵循以下原则：

第一，介绍背景资料时，如果句子的内容是不受时间影响的普遍事实，应使用现在时。如范例中"研究背景"部分。

第二，如果句子的内容是对某种研究趋势的概述，则使用现在完成时。如范例中"研究背景"部分。

第三，叙述作者的工作和结果一般用过去时。如范例中"研究方法和研究结果"部分。

（3）多数期刊对英文摘要字数有要求，一般不宜超过250个实词，要求结构严谨、语义确切、表述简明、不分段落。

（4）不使用非本专业的读者尚难以理解的缩略语、简称、代号，如确有需要使用非同行熟知的缩写，应在缩写符号第一次出现时给出其全称。

（5）应尽量避免使用图、表、化学结构式、数学表达式、角标和希腊文等特殊符号，一般不引用参考文献。

3. 案例分析

范例：Abstract

Climate change is predicted to cause a decline in warm-margin plant populations, but this hypothesis has rarely been tested（研究背景）. Understanding which species and habitats are most likely to be affected is critical for adaptive management and conservation（研究目的）. We monitored the density of 46 populations representing 28 species of arctic-alpine or boreal plants at the southern margin of their ranges in the Rocky Mountains of Montana, USA, between 1988 and 2014 and analysed population trends and relationships to phylogeny and habitat（研究方法）. Marginal populations declined overall during the past two decades;

however, the mean trend for 18 dicot populations was −5.8% per year, but only −0.4% per year for the 28 populations of monocots and pteridophytes. Declines in the size of peripheral populations did not differ significantly among tundra, fen and forest habitats（研究结果）. Results of our study support predicted effects of climate change and suggest that vulnerability may depend on phylogeny or associated anatomical/physiological attributes（研究结论）.（Peter and Crone, 2017）

二、关键词（Keywords）

关键词是作者在完成论文写作后，选出的能表示论文主要内容的词汇。关键词既可以作为文献检索或分类的标识，又是论文主题的浓缩。在提炼关键词时应注意以下几点：

（1）关键词一般是名词或名词词组，可以从论文标题和摘要中选择。

（2）不要使用过于宽泛的词做关键词（如 research, methods, effects, study 等）。

（3）避免使用自定的缩略语、缩写字作为关键词，除非是科学界公认的专有缩写字（如 DNA）。

（4）关键词的数量要适中，一般为 3~8 个词作为关键词。根据不同期刊要求不同，关键词一般以分号、逗号或空格分开。

（5）关键词的首字母一般大写，并且按关键词的首字母顺序排列。

Lesson 55　前言撰写技巧

前言（Introduction）位于正文的起始部分，是 SCI 论文写作最难写的部分之一，同时也是最重要的组成部分。一个好的前言对于论文能否被接收和发表至关重要。因此，作者应该在前言的撰写上多下功夫。本课以"Original Research Papers"类型论文为例，详细介绍前言的写作技巧，并结合具体案例进行分析。

一、前言主要组成部分

"Original Research Papers"类型论文前言由四部分组成,如果能把这四个组成部分说清楚,前言基本上就完成了。

(1) 研究背景。与本研究工作的有关背景简单介绍,也就是作者为什么要做这项研究。

(2) 相关文献总结。这部分是前言中重要组成成分,需要详细来写。主要是概括总结该领域过去和现在研究状况。

(3) 分析现有研究存在问题,提出作者研究内容。通过分析前人研究的局限性,提出作者要研究的内容。在阐述前人研究局限性时,需要客观公正评价别人的工作。

(4) 研究意义。就是针对本研究内容做出一个简单客观的评价。

二、前言写作要求

(1) 尽量准确、清楚且简洁地阐述自己研究领域的基本内容。

(2) 文献要尽量全面客观,特别是对最新的进展和过去经典文献的引用。最好有近三年的新文献。

(3) 不要过分贬低和批评别人的研究工作。给自己的研究工作一个恰如其分的地位,既不要夸大又不能缩小。

(4) 作者在前言中提出自己的研究内容后,习惯提出假设。

(5) 前言中一般不展开讨论。

三、案例分析

通过下面"前言"案例,回答如下问题:

(1) 作者为什么做 UV-B 辐射和氮沉降研究?

(2) UV-B 辐射和氮沉降研究现状。

(3) 作者研究的主要内容及其创新性。

(4) 作者研究的贡献或意义。

(5) 注意这部分写作中时态的变化,归纳"前言"写作中时态运用原则。

案例: Introduction

Atmospheric ozone remains depleted and the annual average ozone loss is approximately 3% globally (Executive summary 2003). Researches have shown that enhanced ultraviolet-B (UV-B) reaching the surface of the earth has very many adverse impacts on plant growth (Jordan, 1996, 2002; Jansen, 2002). When plants are exposed to UV-B stress, they could induce some protective mechanisms. For example, the increases in UV-B absorbing compounds, proline content, and activity of antioxidant enzymes have been reported (Baumbusch et al., 1998; Prochazkova et al., 2001; Saradhi et al., 1995).

In addition to UV-B radiation, human activities have significantly altered the global nitrogen cycle, with the development of industry and agriculture. More and more nitrogen will be imported into the terrestrial ecosystems through nitrogen deposition. In the European livestock and industrialized areas, nitrogen deposition was more than 25 kg hm^{-2} a^{-1} N (Binkley et al., 2000). In the Northeastern United States, the current nitrogen deposition was more 10–20 times than nitrogen in background (Magill et al., 1997). At present, China has been one of three high-nitrogen deposition regions (Li et al., 2003).

Nitrogen is the mineral nutrient needed in largest amounts by plants and it is usually also the limiting factor for plant growth in terrestrial ecosystems (Vitousek and Howarth, 1991), particularly in tundra, boreal as well as alpine ecosystems (Xu et al., 2003). At the same time, nitrogen is also an important constituent of photosynthetic apparatus (Correia et al., 2005). Maximum photosynthetic capacity is strongly regulated by leaf nitrogen concentration (Field and Mooney, 1986). In contrast to UV-B radiation, supplemental nitrogen improved growth and net photosynthesis of plant (Nakajietal., 2001; Keski-Saari and Julkunen-Tiitto, 2003) and reduce production of free radicals in plants (Ramalho et al., 1998). UV-B radiation and nitrogen are expected to increase simultaneously with future changes in global climate. Nitrogen can affect UV-B response in plants (Correiaetal, 2005; Pinto et al., 1999). Previous studies have mainly focused on crop and herb plants, although forests account for over two-thirds of global net primary productivity (NPP), compared with about 11% for agricultural land (Barnes et al., 1998). However, only limited papers have been reported on the combined effects of nitrogen and UV-B radiation on woody plants.（De La Rose et al., 2001, 2003; Lavola et al. ,2003）

Picea asperata is a key species in the southeast of the Qinghai-Tibetan Plateau of China and widely used in reforestation programs at present (Liu, 2002). The paper mainly studies the short-term influence of enhanced UV-B radiation and supplemental nitrogen on photosynthesis and antioxidant defenses of *P. asperata* seedlings under semi-field condition. This will be helpful for understanding of the combined effects on conifer tree species and development of improved plant tolerance toward stressful environmental factors. On the basis of previous study in other species, we hypothesized that (1) both UV-B and nitrogen would affect photosynthesis and antioxidant defenses of *P. asperata* seedlings; and (2) supplemental nitrogen modifies the adverse effects of UV-B on the conifer plants, in order to better understand the responses of woody plant to both enhanced UV-B and to supplemental nitrogen in future.（Yao and Liu, 2007）

Lesson 56　材料与方法撰写技巧

"材料与方法"主要用来介绍试验对象、条件、使用材料、试验设计、参数测定或计算过程、公式推导、模型建立及数据统计方法等。SCI 论文中材料与方法是判断论文创新性、合理性和科学性的主要依据。科学研究的基本要求是研究结果可以重复，材料与方法为他人重复此项研究提供资料。因此，这部分写作要求清楚、准确和详细。下面分别介绍"材料与方法"中每一部分的基本内容和写作要求，并结合案例进行具体分析。（Si et al., 2015）

SCI 论文中"材料与方法"部分常见提纲：

范例1：

Materials and Methods

Plant material and experiment design

UV-B treatments and nitrogen treatments

Pigment analysis

⋮

Statistical analysis

范例2：

Materials and Methods

Study site

Experiment design

Sampling and sample analysis

Data analyses

1. 试验材料

试验材料部分主要介绍试验中所用的材料，要作详细的说明，应写明材料的来源、数量和规格。对于试验用的动物、植物和微生物要正确表明它们的名称，还要写明它们的特征（年龄、性别、遗传学和生理学上的状况）。

2. 试验设计

试验设计部分要详细介绍试验设计，即该试验是如何做的，必须给出详细的细节，让别人能够按照该试验描述进行重复。同时要说明试验处理的重复次数。

范例：Plant material and experiment design

The research was conducted at Hebei University, Baoding, China.

The seedlings of Qi chrysanthemum were obtained from Anguo Chinese herbal medicine planting base, Hebei province, China. The seedlings of the same size were selected based on plant height, and planted into the farmland. Routine field managements were conducted during growth. Fresh flowers were collected when 2/3 of the tubular flowers in the flower head were in bloom.

The harvested flowers were immediately treated with UV-B radiation for 120 min. The experiment included six UV-B radiation intensities: (1) 0 $\mu W\ cm^{-2}$ (UV0); (2) 50 $\mu W\ cm^{-2}$ (UV50); (3) 200 $\mu W\ cm^{-2}$ (UV200); (4) 400 $\mu W\ cm^{-2}$ (UV400); (5) 600 $\mu W\ cm^{-2}$ (UV600) and (6) 800 $\mu W\ cm^{-2}$ (UV800). After UV-B radiation, the samples were put into the incubator (25℃, 80% humidity) for 12 hours. Each treatment had five replications.

UV-B treatments

Enhanced UV-B radiation was produced by UV-B fluorescent lamps (40W, 305 nm, Beijing Electronic Resource Institute, Beijing, China) mounted in metal frames. In the control incubator, UV-B from the lamps was excluded by wrapping the tubes with 0.125 mm polyester film (Chenguang Research Institute of Chemical Industry, China), which transmits UV-A.

3. 参数测定方法

对每一指标的测定方法描述要详略得当、重点突出，所采用的方法必须是公开报道过的方法，需要写明引用的相关文献或者实验指导书，如果在试验过程中对先前已公开出版的方法进行了少量的修改，写作过程中要注明。

范例：Hydrogen Peroxide

Hydrogen peroxide (H_2O_2) content was determined as described by Prochazkova et al. (2001). Sample of 0.5 g was ground with 5 mL cooled acetone in a cold room (10℃). Mixture was filtered with filter paper followed by the addition of 2 mL 5% titanium sulfate and 5 mL ammonium solution to precipitate the titanium-hydrogen peroxide complex. The reaction mixture was centrifuged at 10 000 × g for 10 min. The precipitate was dissolved in 5 mL of 2 mol L^{-1} H_2SO_4 and then recentrifuged. The absorbance of supernatant was measured at 415 nm by spectrophotometer.

4. 数据统计

该部分主要介绍论文中所有数据处理采用什么软件、使用什么方法。

范例：Statistical analysis

All data were subjected to an analysis of variance using the Software Statistical Package for the Social Science (SPSS) version 13.0. Homogeneity of variance was tested using the Levene test prior to analysis. Results were statistically analyzed using one-way ANOVA, followed by Duncan's test to determine whether they were significantly different at the 0.05 probability level.

Lesson 57　结果、讨论和结论撰写技巧

一、结果（Results）

结果是论文的核心，是作者通过试验观测或调查所得出的结果以及各种图像和数据资料。这部分在写作过程中要简明扼要，合理展示数据，证明研究结果与假设一致或相反。这部分在论文中多数情况是单独作为一节来写，有部分期刊要求结果与讨论放在一起来写，即边写边讨论。下面以结果单独作为一节为例介绍这部分写作要求，并结合具体案例进行分析。

结果部分写作要求

（1）结果中图和表要求，一篇论文通常会包含好多数据，这些数据就需要用图或表表示出来，国际期刊要求文中的表一般用三线表，文章中图和表格制作要规范和清楚（见范例1）。

（2）要实事求是地汇报结果或数据，无须加入自己的解释。即使得到的结果与实验不符，也不可不写，而且还应在讨论中加以说明和解释。要根据表或图中统计分析后的数据来写。注意处理间有没有显著差异是根据统计结果来分析的，而不是根据数据的大小（见范例2）。

（3）文字表达应准确、简洁、清楚。应在句子中指出图表所揭示的结论，并把图表的序号放入括号中，让读者清楚知道所描述的结果是哪个图。例如，Enhanced UV-B markedly reduced Chl a, Chl b, Chl (a + b), and Car content。

（4）这部分主要是叙述或总结研究结果，常用一般过去时。

范例1: Figure 1 and Table 1

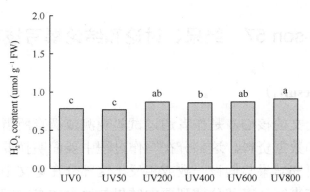

Figure 1 The effects of enhanced UV-B radiation on H_2O_2 content in postharvest flowers of chrysanthemum. UV0, $0\mu W\ cm^{-2}$; UV50, $50\mu W\ cm^{-2}$; UV200, $200\mu W\ cm^{-2}$; UV400, $400\mu W\ cm^{-2}$; UV600, $600\mu W\ cm^{-2}$; UV800, $800\mu W\ cm^{-2}$. The bars with different letters are significantly different from each other ($P < 0.05$). Values are means of five repetitions

Table 1. The effects of enhanced UV-B and supplemental nitrogen on photosynthetic pigment of *Picea asperata*. Values are the mean ± SE of six replicates in column rows and the values in the same column with different letters are significantly different from each other ($P < 0.05$). C, ambient UV-B without supplemental nitrogen (control); N, ambient UV-B with supplemental nitrogen; UV-B, enhanced UV-B without supplemental nitrogen; UV-B + N, enhanced UV-B with supplemental nitrogen.

Treatments	Chl a (mg g^{-1} FW)	Chl b (mg g^{-1} FW)	Chl a/b (mg g^{-1} FW)	Chl (a + b) (mg g^{-1} FW)
C	0.35 ± 0.01 b	0.10 ± 0.00 b	3.50 ± 0.38 a	0.45 ± 0.02 b
N	0.43 ± 0.00 a	0.12 ± 0.00 a	3.58 ± 0.31 a	0.55 ± 0.00 a
UV–B	0.28 ± 0.01 c	0.07 ± 0.01 c	4.00 ± 0.49 a	0.35 ± 0.05 c
UV–B+N	0.30 ± 0.00 c	0.08 ± 0.01 c	3.75 ± 0.24 a	0.37 ± 0.00 c

范例 2：Results（根据范例 1 中 table 1 中数据）

Enhanced UV-B markedly reduced Chl a, Chl b and Chl (a + b) (Table 1). On the other hand, Chl a, Chl b, and Chl (a + b) content of plants grown at ambient UV-B were increased by supplemental nitrogen, whereas supplemental nitrogen did not influence chlorophyll pigment under enhanced UV-B. A parallel change trend in Chl a and Chl b resulted in no significant change in Chl a/b ratio under enhanced UV-B or supplemental nitrogen.

二、讨论（Discussions）

讨论是论文升华部分，也是论文中比较难写的部分之一，讨论的一个重要作用就是突出自己研究的创新性，这就需要和相关的研究进行比较分析。例如，作者研究结果和别人已经发表的结果是否一致，如果一致，是否能够推出某些规律或结论？如果不一致，可能的原因是什么？因此，这部分需要引用大量的参考文献。

1. 讨论部分主要内容

（1）第一段主要概述自己的研究目的和假设。选择要深入讨论分析的问题，对于和前人研究相一致的结果，就不需要再深入讨论，而对于自己研究中新的发现或与相关研究不同的结果要进行重点讨论和分析原因。

（2）指出根据自己的研究结果所得出的结论或推论。

（3）指出本研究所受到的限制及这些限制对研究结果可能产生的影响，并建议进一步的研究题目或方向。

（4）指出自己研究中的理论意义和实际应用价值。

2. 讨论写作注意事项

（1）对结果的解释要重点突出，简洁、清楚，结论需严格客观。

（2）在讨论中，需要特别指出的是要保持和结果的一致性，也就是结果和讨论要一一对应，前后呼应。

（3）讨论部分只对本文结果进行讨论，与同类研究结果进行比较。

（4）对结果的科学意义和实际应用效果的表达要实事求是，不能过分夸大，也不能太谦虚。

（5）时态的运用。回顾研究目的和概述研究结果时，通常使用过去时，阐述由结果得出的推论时，具普遍有效的结论或推论（而不只是在讨论自己的研究结果），并且结果与结论或推论之间的逻辑关系为不受时间影响的事实，通常使用现在时。

3. 案例分析

范例：Discussions

Many studies have shown that enhanced UV-B radiation reduced chlorophyll content and net photosynthetic rates in leaves. The negative effects of enhanced UV-B radiation on chlorophyll content in leaves of lemon grass were found to be dose-dependent (Kumari et al. 2010). Smith et al. (2000) studied the sensitivity of a variety of vegetable crop to enhanced UV-B radiation, and the results showed that chlorophyll content in these plants all had varying degrees of reduction. In this study, the responses of chlorophyll a and chlorophyll b in postharvest flowers to enhanced UV-B radiation were differences. Chlorophyll a content was not affected by UV-B treatments. UV600 and UV800 treatments induced a significant increase in chlorophyll b content, which probably led to the increase in total chlorophyll content under UV600 and UV800. The increased chlorophyll content induced by enhanced UV-B radiation was previously observed in a few researches (Arróniz-Crespo et al. 2011; Sakalauskaité et al. 2013). However, UV radiation in these studies was applied during growth. This study explored the UV-B effects on the chlorophyll content in postharvest flowers. The results showed that appropriate UV-B radiation probably promoted photosynthetic efficiency in postharvest plant organs by the increased chlorophyll content（Si et al., 2015）。

三、结论（Conclusions）

结论是作者对研究的主要发现和成果进行概括总结，让读者对全文的重点有一个深刻的印象。结论应完整、准确、鲜明地表达作者的观点。有的文章也在本部分提出当前研究不足之处，对研究前景和后续工作进行展望。这部分在论文写作中有时单独列为一节，有时放在"讨论"或"结果与讨论"中。

1. 结论写作注意事项

（1）结论要来源于论文，不能编造无法从论文中导出的结论。

（2）要与前言相呼应，不能模棱两可。

（3）不做自我评价。

（4）结论中总结结果语句一般用过去时，对于展望语句用一般现在时或将来时。

（5）常见的结尾句，如 It can be concluded that…, The following conclusions can be drawn from…, The results indicated that…, In conclusions…。

2. 案例分析

范例1：Conclusions

UV-B-effects on biochemical traits in postharvest flowers depended on UV-B radiation levels. <u>The results indicated that optimal UV-B radiation could promote secondary metabolism processes and increase medically active ingredients in postharvest flowers. Proteomic analysis revealed that 19 differentially expressed protein spots were successfully indentified by MALDI-TOF MS. These proteins were mainly involved in photosynthesis, respiration, protein biosynthesis and defence, and secondary metabolism</u> （总结结果，过去时）. <u>This study provides new insights into the responses in postharvest flowers to enhanced UV-B radiation. Further studies are needed to better understand the molecular basis of the UV-B effects on medically active ingredients in postharvest plant organs</u> （一般现在时）（Yao et al., 2015）.

范例2：Conclusions

In conclusion, <u>enhanced UV-B led to a marked decline in wheat yield and protein concentration, and influenced nutritional element concentration of wheat grain. Se supply increased wheat yield and protein concentration, and increased most micro-element concentration in wheat grains, which further improved quality of wheat subjected to UV-B stress to some extent</u> （结果总结，过去时）. <u>Our results confirmed our hypothesis</u> （与前言假设相呼应）（Yao et al., 2013）.

Lesson 58　致谢和参考文献撰写技巧

一、致谢（**Acknowledgements**）

科学研究工作常常需要多方面的指导和帮助才能完成，因此，当科研成果以论文形式发表时，有时需要对他人劳动给予充分地肯定，并郑重地以书面形式表示感谢,其顺序最好按贡献大小排列,而不要按年龄、地位排列名次。致谢的言辞应该恳切,实事求是,恰如其分,而不应浮夸或单纯的客套。致谢的语句要尽量简短。

1. 致谢对象和范围

(1) 在试验过程中给予过技术支持的单位、团体或个人。

(2) 给予经费资助的单位、团体或个人(如国家自然科学基金、省自然科学基金等)。

(3) 在论文选题、构思、试验设计、数据收集及论文撰写过程中提供建设性的意见及提供指导的老师、同事或同学。

(4) 提供过实验材料、仪器,及给予其他方便的人。

2. 案例分析

范例1:Acknowledgements

During this work, the senior author was supported by the National Natural Science Foundation of China (No. 30530630), the Talent Plan of the Chinese Academy of Sciences and Knowledge Innovation Engineering of the Chinese Academy of Sciences(对经费资助单位感谢)(Yao and Liu, 2007).

范例2:Acknowledgements

We are very grateful to G. Glauser (University of Neuchâtel, Switzerland) for the analysis of prenylquinones. We also thank V. Burlat and V. Courdavault for providing the BiFC vectors, M. Rothbart and B. Grimm for the GGR clone, P. Obrdlik for the Y2H plasmids, F. Wüst for his support with the Y2H setup and M. Nater for her phenotyping work(对试验中给予帮助的人感谢). Technical support from M.R. Rodríguez-Goberna and members of the CRAG Services is greatly appreciated. We would finally like to thank M. Meret from the Max Planck Institute of Molecular Plant Physiology in Golm (Germany) for support in the metabolomics sample preparation procedure(对于试验中给予技术支

持的个人和团体表示感谢）。This work was supported by grants from CYTED (Ibercarot-112RT0445), MINECO (BIO2011-23680), AGAUR (2014SGR-1434), ETH Zurich (TH-51 06-1), the EU FP7 contract 245143 (TiMet) and SK grant VEGA 1/0417/13. M.A.R-S. and M.V.B. were supported by MINECO FPI and AGAUR FI fellowships（对给予经费资助单位感谢）(Ruiz-Sola et al., 2016)。

二、参考文献（References）

参考文献是科技论文的重要组成部分，凡在论文中引用前人已发表文献中的观点、数据和材料等均应对它们在正文中出现的地方予以标明并在文末列出参考文献。参考文献的主要作用是反映论文的科学依据，表现作者对他人研究成果的尊重，向读者提供文中引用有关资料的出处，或为了节约篇幅或叙述方便，提供在论文中涉及而没有展开的有关内容的详尽文本。不同期刊对参考文献要求不同，因此，作者在选好期刊后应按期刊要求修改参考文献格式。

1. 参考文献选用原则

（1）尽量选用原始文献，杜绝使用二次文献引用。

（2）尽量选用较新的文献，一般应以引用近三年发表的相关文章为主，这样更能突出本领域的最新研究进展和自己科研的意义和价值。但是，本领域经典的文献一定要引用。

（3）确保文献正确无误（作者姓名，论文题目，期刊或专著名，期刊的年/卷/期或专著的出版年、出版地、出版社、起止页码等）。

（4）避免过多地特别是不必要地引用作者本人的文献。

（5）尽量引用高水平的文献。

2. 文中参考文献的标注方法

（1）根据参考文献在文中出现的先后顺序按阿拉伯数字序号编排。例如：Researches have shown that enhanced ultraviolet-B（UV-B）reaching the surface of the earth has very many adverse impacts on plant growth（1，2）。

（2）用所引用文献中作者姓和文献发表年代来表示。例如：Researches have shown that enhanced ultraviolet-B（UV-B）reaching the surface of the earth has very many adverse impacts on plant growth（Jordan, 1996, 2002; Jansen, 2002）。

3. 参考文献列表编排方法

（1）按作者姓氏字母顺序排列。

（2）按文中参考文献出现的先后顺序编排，即对各参考文献按引用的顺序编排序号。

Lesson 59　期刊选择技巧

论文完成后面临的关键问题就是选择投稿期刊，对于经验丰富的科研工作者，一般在撰写文章时心中就已经有了目标期刊。而对于大学生和研究生，由于缺乏该方面的经验，通常会觉得无从下手。如何选择期刊是一门学问。选择恰当的期刊是论文能够快速发表的一个重要环节。下面将主要介绍期刊选择技巧及注意事项，以便指导作者论文能够快速发表。

一、期刊选择技巧

（1）对自己的文章进行自评，即从创新性和研究意义及技术难度等方面进行大致评估，从而为期刊的选择提供指导。

（2）自己所写文章的研究领域要与期刊所包括的领域相一致。投稿前最好查一下和自己研究相近的论文都在哪些期刊上发表，然后再结合自己论文的质量进行选择。找到期刊主页，查看其"目标和范围"(aim and scope)，以确定稿件是否符合。

（3）明确期刊周期和每年刊载量。期刊的周期一般有：周刊、半月刊、月刊、双月刊、季刊、半年刊和年刊。一般来讲周期短的期刊需要的文章数量大，相对来说刊载多的期刊发表的机会可能多一些。

（4）了解所选 SCI 期刊的审稿周期。一篇文章从投稿到正式发表，不同的期刊效率不同。如何了解不同期刊的审稿周期呢？可以以近期该期刊发表论文的首页了解该信息。例如，若作者想投 Scientific Reports，那么从图 12.1 中红色圆圈标注的地方可以得知，这篇论文是 2015 年 8 月 14 号被该期刊收到，2015 年 11 月 03 号接收，2015 年 12 月 10 号出版。根据这些信息可以判断一篇文章在该期刊上发表大概需要多长时间。

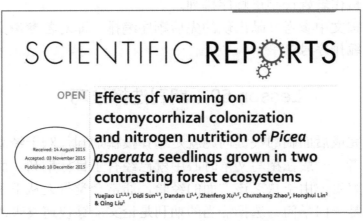

图 12.1　Scientifc Reports 收录论文查询结果

二、期刊选择注意事项

（1）注意所选英文期刊是否收费。期刊收费情况一般在期刊主页中"Author guidelines"中有介绍。发表英文论文费用一般包括版面费和彩色图片印刷费。SCI 数据库中多数期刊是免费的，但如果作者选择论文开放发表或者论文中的图要求彩色打印则需要交纳费用。另外，如果作者索要单篇论文打印本也需要交费。

（2）不同刊物有不同的固定格式和版式特点，期刊选定后要根据期刊要求对论文进行格式上的修改，尤其是参考文献格式。避免因自己的论文格式与所投刊物要求不相符而被退稿。

（3）谨慎辨别，因为有些期刊名称非常相似，甚至有些期刊名称完全相同。因此，网上在线投 SCI 期刊论文时一定要注意 ISSN 号和

Unit 12　SCI 期刊论文写作技巧

期刊主页上的影响因子。例如：图 12.2 和图 12.3 是两个名称完全相同的期刊，但图 12.3 中的期刊是被 SCI 收录的，图 12.2 中的期刊不被 SCI 收录。在此特意指出，希望作者不要上当受骗。

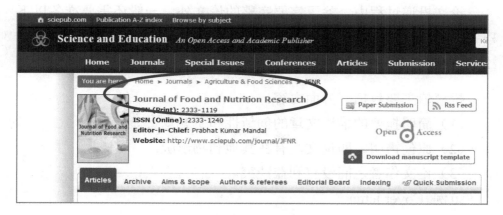

图 12.2　Journal of Food and Nutrition Research 主页界面

图 12.3　Journal of Food and Nutrition Research 主页界面

Lesson 60　SCI 期刊论文投稿时需要的内容

论文投递过程中，除了需要完整的论文外，一般还需要准备以下内容。

一、投稿信（Cover letter）

投稿信主要包括三方面内容：

（1）简明扼要的描述文章的创新性和研究意义。

（2）原创声明，即论文没有在其他刊物出版。

（3）论文作者之间没有利益冲突。

范例: Cover letter

Dear Editor-in-Chief,

I/We am writing to submit my/our manuscript entitled "<title>" for consideration for publication in "<journal name>". "<A brief description of the innovation and significance of the article>".

This manuscript describes novel work and is not under consideration for publication/published by any other journal. All author(s) have approved the manuscript and this submission.

The author(s) certify that there is no conflict of interest with any financial/research/academic organization, with regards to the content/research work discussed in the manuscript.

Thank you for receiving my/our manuscript and considering it for review. Kindly contact me/us at the following address for any future correspondence.

Best Regards,

*Corresponding Author's Name: ×××××

Affiliation: ×××××××, ×××××××

Email Address: ×××@×××.com

Tel. No:
Fax:

二、研究亮点（**Research highlights**）

研究亮点（创新性或新发现）一般提供 3～5 条，每条不超过 85 个字符（包括空格），目前，好多杂志都要求投稿时提供，主要是供评审专家阅读。

范例：

Research highlights

Nutrient and active ingredient contents in flower depended on floral development.

UV-B increased nutrient and active ingredient contents in four stages of flowers.

The best harvest stage of flowers was between the bud and young flower stages.

三、对审稿专家所提意见答复（**Response to reviewer**）

论文修改后再次递交时需要提交对审稿专家所提意见的答复。改文件主要包括以下几方面内容：修改论文的名称、论文的编号、对评审专家和编辑部的感谢及对评审专家和编辑部提出意见的逐条回答等。

范例：

Response to reviewers Editor Journal of Photochemistry and Photobiology B: Biology

Dec 28, 2015

Dear Carpentier

Thank you very much for your letter on Dec 17, 2015, and comments, with regard to our manuscript "article title (No. ****)". We have revised our manuscript according to editor's and reviewers' comments. We believe that the revised manuscript has been improved satisfactorily and hope it will be accepted for publication in Journal of Photochemistry and Photobiology B: Biology. The changes in revised paper were marked with red colour text.

Our responses to reviewers' comments are as follows:

Reviewers' comments

Reviewer #1

1. The authors determined a lot of indices. Therefore, it is need more widely discussed about importance of this indices and UV-B impact on them in the introduction.

Yes, I added some in introduce of revised paper, and discussed UV-B impact on them in the discussions.

2. In the Materials and Methods it is need to describe clearly from fresh or dry mass all analysis was made.

Yes, I added it in revised paper.

Reviewer #2

1. Title is confusing and should to be "Effects of enhanced UV-B radiation on the nutrient and active ingredient contents during the floral development of medicinal chrysanthemum"

Yes, I changed it in revised paper.

Abstract: line 33, remove H_2O_2

Because, H_2O_2 has appeared two times in the abstract, the abbreviation was used in second time.

Thanking again for your kind help and comments and suggestions.

With best regards,

Sincerely yours,

****

References

Part I

Annenberg Learner staff. Evolution and natural selection in ecosystems. [2017-01-22]. http://learner.org/courses/envsci/unit/text.php?unit=4&secNum=8.

Annenberg Learner staff.Energy flow through ecosystems. [2017-01-02] http://learner.org/courses/envsci/ unit/text.php?unit=4&secNum=3.

Boundless. Ecological succession. [2016-12-25]. https://www.boundless.com/biology/textbooks/boundless-biology-textbook/population-and-community-ecology-45/community-ecology-254/ecological-succession-939-12198/.

Boundless. Population size and density. [2016-11-26]. https://www.boundless.com/ biology/textbooks/boundless-biology-textbook/population-and-community-ecology-45/population-demography-249/population-size-and-density-925-12181/.

Boundless. The role of species within communities. [2016-12-20]. https://www.boundless.com/biology/textbooks/boundless-biology-textbook/population-and-community-ecology-45/community-ecology-254/the-role-of-species-within-communities-938-12197/.

Boundless. The study of population dynamics. [2016-12-05]. https://www.boundless.com/biology/textbooks/boundless-biology-textbook/population-and-community-ecology-45/population-demography-249/the-study-of-population-dynamics-927-12183/.

Boundless. Theories of life history. [2016-12-05]. https://www.boundless.com/biology/textbooks/boundless-biology-textbook/population-and-community-ecology-45/life-history-patterns-250/theories-of-life-history-932-12189/.

Camill P. 2010. Global change: An overview. Nature Education Knowledge, 3: 49.

Chapman P M. 2012. Global climate change and risk assessment: invasive species. Integrated Environmental Assessment and Management, 8: 199-200.

Diffen. Abiotic vs Biotic. [2016-11-25]. http://www.diffen.com/difference/Abiotic_vs_Biotic.

Hong S K, Lee J A. 2006. Global environmental changes in terrestrial ecosystems. International issues and strategic solutions: Introduction. Ecology Research, 21: 783-787.

Jørgensen S E. 2011. Ecosystem Ecology. Beijing: Science Press.

McGlynn T P. 2010. Effects of biogeography on community diversity. Nature Education Knowledge, 3: 42.

Molles M. 2000. Ecology: Concepts and Applications. Beijing: Science Press.

Smee D L. 2010. Species with a large impact on community structure. Nature Education Knowledge, 3: 40.

Smith R L, Smith T M. 2003. Elements of Ecology. Fifth edition. Benjamin Cummings, New York.

The free encyclopedia. Ecosystem. [2017-01-20]. https://en.wikipedia.org/wiki/Ecosystem.

The free encyclopedia. Foundation species. [2016-12-12]. https://en.wikipedia.org/wiki/ Foundation_species.

The free encyclopedia. Keystone species. [2016-12-10]. https://en.wikipedia.org/wiki/ Keystone_species.

The free encyclopedia. Species distribution. [2016-11-27]. https://en.wikipedia.org/wiki/ Species_distribution.

Willis K J, Bhagwat S A. 2009. Biodiversity and climate change. Science, 326: 806-807.

Part II

Bennett H, Altenrath C, Woods L, et al. 2017. Interactive effects of temperature and p CO_2 on sponges: from the cradle to the grave. Global Change Biology, 23: 2031-2046.

Black C K, Davis SC, Hudiburg TW, et al. 2017. Elevated CO_2 and temperature increase soil C losses from a soybean-maize ecosystem. Global Change Biology, 23: 435-445.

Blouin M, Karimi B, Mathieu J, et al. 2015. Levels and limits in artificial selection of communities. Ecology Letters, 18: 1040-1048.

Bunn R A, Ramsey PW, Lekberg Y. 2015. Do native and invasive plants differ in their interactions with arbuscularmycorrhizal fungi?A meta-analysis. Journal of Ecology, 103:1547-1556.

Cebrian J. 2015. Energy flows in ecosystems—Relationships between predator and prey biomass

References

are remarkably similar in different ecosystems. Science, 349: 1053-1054.

Currie D J, Venne S. 2017. Climate change is not a major driver of shifts in the geographical distributions of North American birds. Global Ecology and Biogeography, 26: 333-346.

Davis M, Grime J, Thompson K. 2000. Fluctuating resources in plant communities: a general theory of invasibility. Journal of Ecology, 88: 528-534.

Deraison H, Badenhausser I, Loeuille N, et al. 2015. Functional trait diversity across trophic levels determines herbivore impact on plant community biomass. Ecology Letters, 18: 1346-1355.

Gomez-Casanovas N, Hudiburg T W, Bernacchi C, et al. 2016. Nitrogen deposition and greenhouse gas emissions from grasslands: uncertainties and future directions. Global Change Biology, 22: 1348-1360.

Hatton I A, McCann K S, Fryxell J M, et al. 2015. The predator-prey power law: Biomass scaling across terrestrial and aquatic biomes. Science, 349: aac6284.

Hempel S, Götzenberger L, Kühn I, et al. 2013. Mycorrhizas in the Central European flora—relationships with plant life history traits and ecology. Ecology, 94: 1389-1399.

Ikeda D H, Max T L, Allan G J, et al. 2017. Genetically informed ecological niche models improve climate change predictions. Global Change Biology, 23: 164-176.

Johnson D W, Freiwald J, Bernardi G. 2016. Genetic dibersity affects the strength of population regulation in a marine fish. Ecology, 97: 627-639.

Kovach R P, Al-Chokhachy R, Whited D C, et al. 2017. Climate, invasive species and land use drive population dynamics of a cold-water specialist. Journal of Applied Ecology, 54: 638-647.

Küster E C, Kühn I, Bruelheide H, et al. 2008. Trait interactions help explain plant invasion success in the German flora.Journal of Ecology, 96: 860-868.

Lafferty K D, DeLeo G, Briggs C J, et al. 2015. A general consumer-resource population model. Science, 349: 854-857.

Lewontin R C. 1970. The units of selection. Annual Review of Ecology & Systematics,1: 1-18.

Ma C H, Chu J Z, Shi X F, et al. 2016. Effects of enhanced UV-B radiation on the nutritional and active ingredient contents during the floral development of medicinal chrysanthemum. Journal of Photochemistry & Photobiology, B: Biology, 158: 228-234.

Mallon C A, Poly F, Roux X L, et al. 2015. Resource pulses can alleviate the biodiversity-invasion relationship in soil microbial communities. Ecology, 96: 915-926.

Mcelroy D, O'Gorman E J, Schneider F D, et al. 2015. Size-balanced community reorganization in response to nutrients and warming. Global Change Biology, 21: 3971-3981.

Melles S J, Fortin M J, Lindsay K E, et al. 2011. Expanding northward: influence of climate change, forest connectivity, and population processes on a threatened species' range shift. Global Change Biology, 17: 17-31.

Menzel A, Hempel S, Klotz S, et al. 2017. Mycorrhizal status helps explain invasion success of alien plant species. Ecology, 98: 92-102

Pacifici M, Visconti P, Butchart S H M, et al. 2017. Species' traits influenced their response to recent climate change. Nature Climate Change, 7: 205-208.

Penn A. 2003. Modelling artificial ecosystem selection: a preliminary investigation//Banzhaf W, Christaller T, Dittrich P, et al. Advances in Artificial Life, Lecture Notes In Artificial Intelligence. Berlin, Germany: Springer-Verlag.

Pringle A, Bever J D, Gardes M, et al. 2009.Mycorrhizal symbioses and plant invasions.Annual Review of Ecology, Evolution, and Systematics, 40: 699-715.

Ruiz-Benito P, Gómez-Aparicio L, Paquette A, et al. 2014. Diversity increases carbon storage and tree productivity in Spanish forests. Global Ecology and Biogeography, 23: 311-322.

Schultz C B, Pe'er B G, Damiani C, et al. 2017. Does movement behaviour predict population densities? A test with 25 butterfly species. Journal of Animal Ecology, 86: 384-393.

Steudel B, Hector A, Friedl T, et al. 2012. Biodiversity effects on ecosystem functioning change along environmental stress gradients. Ecology Letters, 15: 1397-1405.

Swenson W, Wilson D S, Elias R. 2000. Artificial ecosystem selection. Proceedings of the National Academy of Sciences, USA, 97: 9110-9114.

Taylor M K, Lankau R A, Wurzburger N. 2016. Mycorrhizal associations of trees have different indirect effects on organic matter decomposition. Journal of Ecology, 104: 1576-1584.

Thom D, Rammer W, Seidl R. 2017. Disturbances catalyze the adaptation of forest ecosystems to changing climate conditions. Global Change Biology, 23: 269-282.

Thompson R M, Beardall J, Beringer J, et al. 2013. Means and extremes: building variability into community-level climate change experiments. Ecology Letters, 16: 799-806.

References

Tilman D. 2004. Niche tradeoffs, neutrality, and community structure: a stochastic theory of resource competition, invasion, and community assembly. Proceedings of the National Academy of Sciences USA 101:10854-10861.

van der Kooi C, Reich M, Löw M, et al. 2016. Growth and yield stimulation under elevated CO_2 and drought: A meta-analysis on crops. Environmental and Experimental Botany, 122: 150-157.

Vayreda J, Martinez-Vilalta J, Gracia M, et al. 2016. Anthropogenic-driven rapid shifts in tree distribution lead to increased dominance of broadleaf species. Global Change Biology, 22: 3984-3995.

Vromman D, Lefèvre I, Šiejkovec Z, et al. 2016. Salinity influences arsenic resistance in the xerohalophyte *Atriplexat acamensis* Phil. Environmental and Experimental Botany, 126: 32-43.

Yasuhara M, Danovaro. 2016. Temperature impacts on deep-sea biodiversity. Biological Reviews. 91: 275-287.

Zander A, Bersier L F, Gray S M. 2017. Effects of temperature variability on community structure in a natural microbial food web. Global Change Biology, 23: 56-67.

Part Ⅲ

穆蕴秋，江晓原. 2016. 为何中国学者和媒体对影响影子公式的表述普遍错误？出版发行研究，9: 9-11.

张峰，武玉珍，张桂萍，等. 2006. 生态学研究中常见的统计学问题分析. 植物生态学报, 30: 361-364.

Li Y J, Sun D D, Li D D, et al. 2015. Effects of warming on ectomycorrhizal colonization and nitrogen nutrition of *Picea asperata* seedlings grown in two contrasting forest ecosystems. Scientific Reports, 5: 1-10.

Peter L, Crone E E. 2017. Arctic and boreal plant species decline at their southern range limits in the Rocky Mountains. Ecology Letters, 20: 166-174.

Pielou E C. 1985. Mathematical Ecology (2nd edition), USA, New York: Wiley-Interscience.

Ruiz-Sola M A, Coman D, Beck G, et al. 2016. Arabidopsis GERANYLGERANYL DIPHOSPHATE SYNTHASE 11 is a hub isozyme required for the production of most photosynthesis-related isoprenoids. New Phytologist, 209: 252-264.

Si C, Yao X Q, He X L, et al. 2015. Effects of enhanced UV-B radiation on biochemical traits in postharvest flowers of medicinal chrysanthemum. Photochemistry and Photobiology, 91: 845-850.

Yao X Q, Chu J Z, He X L, et al. 2013. Effects of selenium on agronomical characters of winter wheat exposed to enhanced ultraviolet-B. Ecotoxicology and Environmental Safety, 92: 320-326.

Yao X Q, Chu J Z, Ma C H, et al. 2015. Biochemical traits and proteomic changes in postharvest flowers of medicinal chrysanthemum exposed to enhanced UV-B radiation. Journal of Photochemistry and Photobiology B: Biology, 149: 272-279.

Yao X Q, Liu Q. 2007. Changes in photosynthesis and antioxidant defenses of *Picea asperata* seedlings to enhanced ultraviolet-B and to nitrogen supply. Physiologia Plantarum, 129: 364-374.

Appendix Ⅰ 常见生态学专业词汇

A

abiotic environment 非生物环境
abundance 多度
acid rain 酸雨
adaptation 适应
aestivation 夏眠
after-ripening 后熟
age distribution 年龄分布
age-specific life table 特定年龄生命表
age structure 年龄结构
agroecology 农业生态学
agroecosystem 农业生态系统
air pollution 空气污染
allelochemical 化感物质
allelopathy 化感作用
Allen's rule 阿伦法则
altitudinal belt 垂直分布带
amount of rainfall 降雨量
annual ring 年轮
antagonism 拮抗作用；拮抗现象
arid 干旱
artificial forest 人工林
asexual reproduction 无性生殖
autecology 个体生态学
autotroph 自养生物
auxin 生长激素
available nitrogen 有效（性）氮

B

Bergman's rule 贝格曼规律
biocoenosis 生物群落
bioconcentration 生物浓缩
biodiversity 生物多样性
biogeochemical cycle 生物地球化学循环

biological control   生物防治
biological enrichment   生物富集
biological oxygen demand   生化需氧量

biomass   生物量
biome   生物群系
biosphere   生物圈
biotic environment   生物环境

C

carnivore   食肉动物
carrying capacity   环境容纳量
chemical oxygen demand (COD)   化学耗氧量
climax   顶极群落
clumped distribution   聚集分布
coevolution   协同进化
colonization   定居，建群
community   群落

community ecology   群落生态学
companion species   伴生种
competition   竞争
competition coefficient   竞争系数
constancy   恒有度
constructive species   建群种
convergent adaptation   趋同适应
coverage   盖度

D

decomposer   分解者
decomposition   分解作用
desert   荒漠
desertification   沙漠化
density-dependent   密度制约
density effect   密度效应
density-independent   非密度制约
density ratio   密度比

development   发育
directional selection   定向选择
dominance   优势度
dominant species   优势种
dynamic-composite life table   动态混合生命表
dynamic life table   动态生命表

E

ecology   生态学
ecological amplitude   生态幅
ecological density   生态密度

ecological dominance   生态优势
ecosystem ecology   生态系统生态学

ecological environment 生态环境
ecological factor 生态因子
ecosystem 生态系统
ecotype 生态型
ecological agriculture 生态农业
ecological succession 演替
ecological sustainability 生态持续性

ecotone 群落交错区（生态交错区）
edge effect 边缘效应
emigration 迁出
endoderm 恒温动物
energy flowing 能量流
eutrophication 富营养化
evolution ecology 进化生态学
exponential growth 指数增长

F

farmland ecosystem 农田生态系统
fauna 动物群
feedback 反馈作用
fitness 适合度
flora 植物群
fluctuation 波动
food chain 食物链

food chain structure 食物链结构
food web 食物网
frequency 频度
freshwater ecology 淡水生态学
freshwater ecosystem 淡水生态系统

G

gap 缺口
gene flow 基因流
gene pool 基因库
geometric growth 几何级数增长
global ecology 全球生态

gradient hypothesis 梯度假说
greenhouse effect 温室效应
green manure 绿肥
gross primary production 总初级生产力

H

habitat 生境
hard wood 阔叶木
hardwood forest 阔叶林
height 高度
herbivore 食草动物

heterogeneity 异质性
heterotrophic succession 异养演替
heterotroph 异养生物
herbicide 除草剂
hibernation 冬眠

homogeneity 同质性
human demography 人口统计学
human ecology 人类生态学
humidity 湿度
humus 腐殖质

hydric 水生的
hydroarch succession 水生演替
hydrobiology 水生生物学
hydrophytie 水生植物

I

immigration 迁入
inbreeding 近亲交配
incidental species 偶生种
individual 个体
innate behavio(u)r 先天行为
inoculation 接种
insect pollination 虫媒授粉
insecticide 杀虫剂

intensity 强度
interspecific relationship 种间关系
intertidal zone 潮间带
intraspecific relationship 种内关系
invasion 侵入
invasive species 入侵物种
island ecology 岛屿生态学
isolation 隔离；分离

K

keystone species 关键种

k-strategist k 对策略

L

land use 土地利用
landscape ecology 景观生态学
larva 幼虫
law of the minimum 最小因子法则

law of tolerance 耐受性法则
leached layer 淋溶层
leaf area index 叶面积指数
life history 生活史

M

macroclimate 大气候
macroelement 大量元素
macroplankton 大型浮游生物
marginal distribution 边缘分布

marginal habitat 边缘栖所
marine 海洋
marine ecology 海洋生态学
marine ecosystem 海洋生态系统

Appendix

mating behavio(u)r 交配行为
mating season 交配季
maximum natality 最大出生率
mesophyll 叶肉
metabolism 代谢作用
microbe 微生物
microclimate 小气候
microcommunity 小群落
microecosystem 微生态系统
microelement 微量元素
micro-environment 微环境
migration 迁徙
mineral 矿物（质）
mineral cycling 矿物质循环
molecular ecology 分子生态学
monoclimax theory 单元演替顶级
mortality 死亡率
mortality curve 死亡曲线

N

natality 出生率
native species 自生种；本地种
natural control 自然防治
natural enemy 天敌
net primary production 净初级生产力
negative feedback 负反馈
niche 生态位
nitrogen cycle 氮循环
nitrogen-fixing bacteria 固氮细菌
nodule bacteria 根瘤菌

O

occurrence 出现；分布
omnivore 杂食动物
oxidation 氧化作用
ozone layer 臭氧层

P

parasite 寄生生物
parasitism 寄生
patch 斑状群；斑块
patchiness 斑块性
perennial grass 多年生禾草
perennial herb 多年生草本
perennial plant 多年生植物
phenology 物候学
phenotype 表（现）型
photoperiod 光周期
photorespiration 光呼吸
photosynthesis 光合作用
phototropism 向光性
phycology 藻类学

phyplankton 浮游植物
physiological ecology 生理生态学
phytochrom 植物色素
pioneer community 先锋群落
plankon 浮游生物
pollination 授粉；传粉
pollinator 授粉者；传粉者
pollution 污染
polyclimax theory 多元顶级理论

population 种群
population ecology 种群生态学
positive feedback 正反馈
predation 捕食
predator 捕食者
prevail climax 优势顶级
prey 猎物
primary production 初级生产
primary succession 初级演替

R

radioactive isotope 放射性同位素
radioactive pollution 放射性污染
rain forest 雨林
random distribution 随机分布
rare species 稀有种
recycling 再循环
relative frequency 相对频度
reproduction 生殖
respiration 呼吸作用

restoration ecology 恢复生态学
richness 丰度
rock type 岩石型
rock vegetation 岩生植被
rotation system 轮作系统
rotational grazing 轮牧
r-strategist r 对策略者
runoff 径流

S

sample size 样本数
sample survey 取样调查
sampling 取样
sampling distribution 样本分布
sampling plot 取样小区
scrub 灌丛
secondary metabolites 次生代谢物质
secondary production 次级生产力

secondary succession 次生演替
selective fertilization 选择受精
sewage 污水
sexual reproduction 有性生殖
similarity 相似度
Simpson's diversity index 辛普森多样性指数
sludge 污泥

social behavio(u)r 社会行为
softwood 针叶材
softwood forest 针叶林
soil microorganism 土壤微生物
soil mineral 土壤矿物
soil moisture 土壤水分
spatial pattern 空间格局
species diversity 物种多样性
specific leaf area 面积比叶

spore 孢子
sporophyll 孢子叶
static life table 静态生命表
succession 演替
survivorship curve 存活曲线
survival rate 存活率
subalpine 亚高山
symbiosis 共生

T

taxa 分类群
terrestrial ecology 陆地生态学
terrestrial plant 陆生植物
tidal zone 潮间带
tide 潮汐
time-specific life table 特定时间生命表

tornado 龙卷风(北美中、西部的)
trace element 微量元素
trophic level 营养级
trophic relationship 营养联系
tundra 冻原
typhoon 台风

U

ultraviolet radiation 紫外辐射
underground runoff 地下径流
underground water 地下水

uniform distribution 均匀分布
Urban ecology 城市生态学

V

vegetation 植被
vegetation type 植被型
vegetative propagation 营养生殖

vernalization 春化作用
vertical migration 垂直迁移
virgin forest 原始林

W

wasteland 荒地
water bloom 藻华，水花
water cycle 水循环
water level 水位
water quality 水质

waters 水域
weathering 风化作用
windbreak 防风林
woodland 林地

X

xerarch succession 旱生演替
xerophyte 旱生植物

xylem 木质部

Z

zooplankton 浮游动物
zoobenthos 底栖动物

zonal distribution 带状分布
zonal vegetation 带状植被

Appendix

Appendix II 生态学学科 SCI 收录期刊简介

序号	期刊名	大类学科	小类学科	影响因子	分区	出版周期
1	Annual Review of Ecology Evolution and Systematics	生物	生态学	9.352	1区	年刊
2	Ecological Monographs	环境科学与生态学	生态学	8.037	1区	季刊
3	Ecology Letters	环境科学与生态学	生态学	10.772	1区	月刊
4	Global Change Biology	环境科学与生态学	生态学	8.444	1区	半月刊
5	Methods in Ecology and Evolution	环境科学与生态学	生态学	6.344	1区	月刊
6	Advances in Ecological Research	环境科学与生态学	生态学	3.92	2区	不规律
7	Diversity and Distributions	环境科学与生态学	生态学	4.566	2区	半月刊
8	Ecology	环境科学与生态学	生态学	4.733	2区	月刊
9	Ecosystems	环境科学与生态学	生态学	3.751	2区	半月刊
10	Functional Ecology	环境科学与生态学	生态学	5.21	2区	半月刊
11	Journal of Applied Ecology	环境科学与生态学	生态学	5.196	2区	半月刊
12	Oikos	环境科学与生态学	生态学	3.586	2区	月刊
13	Basic and Applied Ecology	环境科学与生态学	生态学	1.836	3区	半月刊
14	Behavioral Ecology and Sociobiology	生物	生态学	2.382	3区	月刊
15	Biotropica	环境科学与生态学	生态学	1.944	3区	季刊
16	BMC Ecology	环境科学与生态学	生态学	2.724	3区	

续表

序号	期刊名	大类学科	小类学科	影响因子	分区	出版周期
17	Ecological Complexity	环境科学与生态学	生态学	1.797	3区	季刊
18	Ecological Modelling	环境科学与生态学	生态学	2.275	3区	半月刊
19	Ecology and Evolution	生物	生态学	2.537	3区	月刊
20	Ecology and Society	环境科学与生态学	生态学	2.89	3区	季刊
21	Ecosphere	环境科学与生态学	生态学	2.287	3区	月刊
22	Freshwater Science	生物	生态学	2.433	3区	季刊
23	Journal for Nature Conservation	环境科学与生态学	生态学	2.22	3区	季刊
24	Oecologia	环境科学与生态学	生态学	2.902	3区	半月刊
25	Restoration Ecology	环境科学与生态学	生态学	1.891	3区	季刊
26	Acta Amazonica	环境科学与生态学	生态学	0.408	4区	季刊
27	Acta Oecologica-International Journal of Ecology	环境科学与生态学	生态学	1.42	4区	半月刊
28	African Journal of Ecology	环境科学与生态学	生态学	0.875	4区	季刊
29	American Midland Naturalist	环境科学与生态学	生态学	0.592	4区	季刊
30	Austral Ecology	环境科学与生态学	生态学	1.598	4区	半月刊
31	Community Ecology	环境科学与生态学	生态学	1.019	4区	不规律
32	Contemporary Problems of Ecology	环境科学与生态学	生态学	0.259	4区	半月刊
33	Ecological Informatics	环境科学与生态学	生态学	1.683	4区	半月刊
34	Ecological Research	环境科学与生态学	生态学	1.338	4区	半月刊
35	Ecoscience	环境科学与生态学	生态学	0.595	4区	季刊

Appendix

续表

序号	期刊名	大类学科	小类学科	影响因子	分区	出版周期
36	Ecotropica	环境科学与生态学	生态学	0.35	4区	半年刊
37	Ekoloji	环境科学与生态学	生态学	0.592	4区	季刊
38	Evolutionary Ecology Research	生物	生态学	0.585	4区	半月刊
39	Journal of Tropical Ecology	环境科学与生态学	生态学	0.975	4区	半月刊
40	New Zealand Journal of Ecology	环境科学与生态学	生态学	1.247	4区	半月刊
41	Northwest Science	环境科学与生态学	生态学	0.412	4区	季刊
42	Polish Journal of Ecology	环境科学与生态学	生态学	0.5	4区	季刊
43	Population Ecology	环境科学与生态学	生态学	1.698	4区	一年三期
44	Rangeland Journal	环境科学与生态学	生态学	1.194	4区	半年刊
45	Revue d Ecologie-la Terre et la Vie	环境科学与生态学	生态学	0.164	4区	季刊
46	Russian Journal of Ecology	环境科学与生态学	生态学	0.456	4区	半月刊
47	Southwestern Naturalist	环境科学与生态学	生态学	0.255	4区	季刊
48	Theoretical Ecology	环境科学与生态学	生态学	2.085	4区	季刊
49	Tropical Ecology	环境科学与生态学	生态学	1.169	4区	半年刊

数据来源：2016 年